IMPORTANT NOTICE

T.F.H. Publications, Inc. is the largest publisher of dog books in the world. As such we are constantly aware of popular titles in every language. Germany is famous for its excellent dog books and the leading publisher there is Kynos Verlag. They published a pair of best sellers entitled **KAMPFHUNDE I** and **KAMPFHUNDE II** written by Dr. Dieter Fleig. Dr. Fleig has an important international reputation as an authority on dogs. When we read these two books a few years ago we were concerned about Dr. Fleig's opinions about dogs and people. We were even upset about the gory details of dogs fighting dogs, dogs fighting bulls, dogs fighting rats and even dogs fighting people! Then we realized that even today there are terribly bloody fights between men, wrestling men between men and women between women. We realized that there are still war dogs, guard dogs, police dogs and hunting dogs, all of which are trained for specific dangerous tasks. But this still was not a strong enough argument for us to publish these two books and we procrastinated for two years.

Finally, after discussions with booksellers, librarians, pet shop owners and dog authorities, we were convinced to publish these two books. The argument to which we succumbed was: *Because you publish a book about wars (The Civil War, World War II, etc.), doesn't mean you advocate wars. It only means you owe history a debt to record the truth so others might benefit by the mistakes of the past.*

We have changed the names of the books from *Kampfhunde I* and *Kampfhunde II*, which literally means *Fighting Dog I* and *Fighting Dogs II*, to more accurately describe the two volumes. Our TS-270 has been titled **THE HISTORY OF FIGHTING DOGS** while TS-271 is called **FIGHTING DOG BREEDS.**

Dr. Dieter Fleig

Translator: William Charlton

THE HISTORY OF FIGHTING DOGS

This book was originally published in the German language by Kynos Verlag, D-5537 Muerlenbach/Eifel, Germany, under the title **KAMPFHUNDE I**. It was followed by a second book in the series **KAMPFHUNDE II**. The author, Dr. Dieter Fleig is considered one of the world's authorities on fighting dogs. He is also the publisher of the Kynos Verlag. Kynos Verlag owns the copyright on the German language edition; TFH owns the copyright on the translation.

Published in 1996 by T.F.H. Publications, Inc.
Manufactured in Neptune, N.J. by T.F.H. at 211 West Sylvania Avenue, Neptune, N.J. 07753 USA

© 1996 by T.F.H. Publications, Inc.

Distributed in the UNITED STATES to the Pet Trade by T.F.H. Publications, Inc., One T.F.H. Plaza, Neptune City, NJ 07753; distributed in the UNITED STATES to the Bookstore and Library Trade by National Book Network, Inc. 4720 Boston Way, Lanham MD 20706; in CANADA to the Pet Trade by H & L Pet Supplies Inc., 27 Kingston Crescent, Kitchener, Ontario N2B 2T6; Rolf C. Hagen Inc., 3225 Sartelon St. Laurent-Montreal Quebec H4R 1E8; in CANADA to the Book Trade by Vanwell Publishing Ltd., 1 Northrup Crescent, St. Catharines, Ontario L2M 6P5 ; in ENGLAND by T.F.H. Publications, PO Box 15, Waterlooville PO7 6BQ; in AUSTRALIA AND THE SOUTH PACIFIC by T.F.H. (Australia), Pty. Ltd., Box 149, Brookvale 2100 N.S.W., Australia; in NEW ZEALAND by Brooklands Aquarium Ltd. 5 McGiven Drive, New Plymouth, RD1 New Zealand; in Japan by T.F.H. Publications, Japan—Jiro Tsuda, 10-12-3 Ohjidai, Sakura, Chiba 285, Japan; in SOUTH AFRICA by Lopis (Pty) Ltd., P.O. Box 39127, Booysens, 2016, Johannesburg, South Africa. Published by T.F.H. Publications, Inc.

MANUFACTURED IN THE
UNITED STATES OF AMERICA
BY T.F.H. PUBLICATIONS, INC.

CONTENTS

THE TRAIL OF BLOOD 9
  A Preface by Ulrich Klever

I. FOREWORD 11

II. INTRODUCTION 13
  The Development of the Dog into a Specialist, 13; Classification by Carl Linnaeus; Classification by Stonehenge in British Rural Sports (1856); The Origin of the Fighting Dog Breeds, 16

III. THE PURPOSES OF FIGHTING DOGS 20
  War Dogs, 20; The Hunting of Dangerous Game, 25; Fights against Bears and Lions, 48; Fights against Bulls, 60; Dogs fighting Dogs, 82; Fights with the Badger, 100; Fights against Rats, 105; Fights against Apes, Monkeys, Opossums, Pigs, Horses and Donkeys, 112; Fights against Man, 121; The Banning of Animal Fights, 125

SUGGESTED READING 135

INDEX 136

THE TRAIL OF BLOOD
A PREFACE BY ULRICH KLEVER

Aggressiveness and fearlessness in the face of death, these are the traits that we prize in these parts in the breeds of fighting dogs. Therefore, stories from their history are glorified and dogs of unconditional keenness and ferocity are always in demand. In this connection these ideas of the real man and his brave dog are of course only a romantic fantasy, since who in these modern times can afford to live with a dog that leaves behind a trail of blood whenever it encounters others (= enemies).

This dream of faithful, four-legged death as a companion haunts the minds of men. This is shown by the increasing number of dogs from the Bull Terrier to the Mastino, which live in or with families, and the considerably larger, unknown number of those for whom a fighting dog is the dream dog, even if the reality of wife, surroundings, and the like allows "only" a Poodle or Basset Hound or no dog at all.

This book is equally as important for the keeper of fighting dogs as for those who dream of keeping fighting dogs. It shows the true history of those breeds which were bred systematically by humans mostly for profit and out of a misguided sense of sportsmanship. They were bred for the sole purpose of fighting to the death. The reader can learn the outcomes of these fights in shocking documentary descriptions. The more you love dogs, the stronger nerves you will need. The lecture, however, helps us to understand and handle the temperaments and types of these breeds. Particularly with these breeds, history is important, because historical conditions are what shaped them. Apart from war and hunting, it was a perverted spirit of sportsmanship. Thus, this book is also a commentary of the British national character.

The author has presented the most precise canine history that I am aware of, and I am well read in several languages in the relevant literature. Everything is thoroughly investigated and documented here, the illustrations are exemplary and unique in their completeness. These illustrations have the same, if not greater, documentary value than the text.

This is possible only because of the author's large collection, which he assembled over many years.

That a great deal of work and commitment go into a book of this kind, I know all too well as an author. All this would not be enough, if the author did not have extensive experience over many years with perhaps the most aggressive of the breeds of fighting dogs and — you will notice this repeatedly as you read — a deep love of the dog.

I. FOREWORD

It is an ancient dream of mankind, the dream of the incorruptible, faithful four-legged companion. The old gods, with fearless, valiant dogs at their sides — the path to the underworld, separated from the mortal realm by incorruptible, powerful guardians, these old myths reflect an ancient yearning of man. His brave dog protects him from his enemies. The dog fills him with pride to have exclusive power over this life's companion that is so dangerous to all other people and animals. His faithful, four-legged companion turns him into a "superman" — like a god.

For as long as the dog has accompanied humans to live with, humans have enjoyed the protection provided by the dog. If at first it was only the barking with which the dog announces strangers, humans soon recognized that the dog could be more than just a warner, it could be a protector. The flashing, powerful bite of his dog was a valuable weapon for him in his constant battle against the hostile world around him. Companion on the hunt, protector of the herds, guardian of house and family, the dog became the first and most important domestic animal of humans. It has often been acclaimed as man's best friend.

Over millennia, the valiant dog has accompanied humans. It traveled with warriors to foreign lands as a frightening offensive weapon; it protected the wagons, villages, and huts of the residents against intruders, and defended women and children with its powerful jaws. The dog guarded its master's herds against wild predators and became the best hunting companion against dangerous game. If you took all the legs of the dogs that died protecting the lives and property of their masters, and heaped them in a great pile, a gigantic mountain of bones would tower over the landscape, a memorial to the faithfulness and dependability of our dog.

This book is an homage to the brave dog, which gave and continues to give its life to serve its master. Fighting dogs are surely the most magnificent representatives of their species. They deserve our recognition, admiration, and understanding. Accompany me as we travel through the millennia of the mutual past of human and fighting dog until we come to the splendid modern descendents of the fighting dog.

II. INTRODUCTION

1. THE DEVELOPMENT OF THE DOG INTO A SPECIALIST

Dog breeding in its earliest stages, as soon as it was carried out systematically, always had the goal of turning the single dog into a specialist through deliberate selection. We must keep in mind, however, that the purpose, the task of the dog, was always emphasized: the good sheepdog, the fast hunting dog, the agile tracking dog, the fearless fighting dog, the dependable watchdog for house and farm, the small, confiding playmate for the children. We must not overlook here that these tasks place demands on the outward form, on the anatomy of the dog. A dog that is useful for hunting fast rabbits or gazelles on flat and open terrain requires a certain body size, long legs, and a deep, but narrow-ribbed chest. The tracking dog in a heavily forested region, full of underbrush, from it we expect a dense, rugged coat, it should not have too long legs and should hunt by scenting. The fighting dog had an imposing outward form and powerful jaws to instill fear and terror. The fight against a bull demanded a dog that offered as little surface area as possible to the bull's horn as it crept low to the ground toward its opponent, and through the power and form of its jaws could grip the bull's head without being tossed off.

Although the primary goal of dog breeding was always directed toward a specific task, the task at hand also demanded specific basic anatomical traits, so that typical anatomical and temperamental features were defined. We cannot and must not equate this with the modern breeding of pedigree dogs, where, laughably, the absence of a single tooth can lead to exclusion from breeding. Nonetheless, old pictures repeatedly show us uniform breed types, such as for Greyhounds, fighting dogs, watchdogs, lapdogs, so that we can already speak of basic breed characters in early historical times, and even of an anatomical breed type.

Today there is no longer a serious discussion about the origin of our dogs. All reputable scholars concur that the canids, the jackal and the fox, that were formerly looked upon as potential ancestors of our domestic dog, are no longer given serious consideration. The sole ancestor is the wolf, with its numerous races. Depending on the geographic factors, we find wolves in very variable sizes and forms.

Some scholars see the beginnings of our domestic dog in Tomarctus, a wolflike ancestral dog, as early as between the Paleolithic and the Neolithic Ages, about 15,000 years ago. A discussion of this question must remain beyond the scope of this book. We now know with certainty, however, that humans have engaged in breeding dogs for millennia and in the process have produced a multitude of breeds. It has not been proved unambiguously at what point planning entered the picture and to what extent chance alone operated. From finds and excavations, however, we can say with certainty that dogs that were very similar to our modern dog breeds lived about 3500 B.C. Thus, specialized dogs for specific tasks have existed for more than 5500 years. A famous British dog scholar (E. C. Ash) claims that if you were to take dogs from the old graves, bring them to life again, and put them in the show ring alongside our modern breeds, our judges would have a hard time in handing out the prizes correctly. From murals, excavations, and bones we can say that more than 5000 years ago there already existed Greyhounds, hunting dogs, fighting dogs, and lapdogs in the typical body forms of modern times.

We may assume that the development of individual breeds took place in narrow geographic units, corresponding to the performance required in these regions. Certainly in these regional populations inbreeding regularly occurred, accompanied by the strict culling of defective offspring. Unusable or defective offspring were not fed. Only those dogs that met the breeding goals were fed, cared for, and used appropriately.

We must stress here, however, that beside the systematic breeding for specific traits, there also doubtlessly existed a host of nameless mongrels, approximately comparable to our modern barnyard cat population. We must not underestimate the significance of these non-pedigree, accidental products of breeding, because these mongrels made up the overwhelming majority of the dog population. Dogs must also have been introduced routinely into systematic breeding from this reservoir, when they corresponded to the requirements in temperament and appearance. On the other hand, flawed products of systematic breeding returned to this general melting pot.

In excavations in the mid-twentieth century, the correctness of the above representations was broadly confirmed. Excavations in Barsbeck (a village in Schleswig-Holstein, Germany), a site from about the time of the birth of Christ, as well as in Feddersen-Wierde, near Bremerhaven, Germany, from the period between about 50 B.C. to 450 A.D., yielded finds of numerous dog skulls. These skulls ranged from Doberman Pinscher/small Poodle size up to Mastiff size. The frequencies of the skull sizes are uniformly distributed, with no particular size present in greater frequency, so that they do not provide evidence of the existence of distinct breeds. Here we have found the wide sea of non-pedigree mongrels; in contrast, the few purebred dogs scarcely turn up in the finds. On the other hand, we have sufficient historically accurate documentation — including written and pictorial evidence, as well as skeletal finds — to prove the existence of clearly defined breeds in early historical times. We will return to this subject in the discussions of the individual breeds of fighting dog.

It is clear that man, by developing breeds, sought to breed specialists. The initial criteria established in the process are decisive not only for the anatomical structure of these breeds but for the psychological traits of the breed as well. The selection for performance, complemented by the breeding for suitable body forms, leads to the formation of breeds and thereby to the division of labor among dogs. I cannot emphasize strongly enough that the breeds created in this way required constant human care and planning. Had these been neglected, the breed would soon have slipped back into the sea of anonymous and non-pedigree mongrels.

To conclude this topic, it would be interesting to look at several early attempts to classify the old dog breeds. It is particularly interesting for us that the fighting dog is always mentioned. Several examples are provided.

II. INTRODUCTION

I. **Roman classification**
1. *Canes villatici* — domestic dogs
2. *Canes pastorales pecuarii* — sheep dogs
3. *Canes venatici* — hunting dogs
4. *Canes pugnaces* or *bellicosi* — fighting dogs
5. *Canes nares sagaces* — scenting dogs
6. *Canes pedibus celeres* — Greyhounds

II. **Classification by Carl Linnaeus (1756) (Swedish naturalist)**
1. *Canis domesticus* — domestic dog
2. *Canis jagax* — hunting dog
3. *Canis graius* — Greyhound
4. *Canis mastinus* — Mastiff — Molossus
5. *Canis aquaticus* — water dog
6. *Canis melitaeus* — lapdog
7. *Canis aegypticus* — Egyptian dog = Greyhound
8. *Canis fricatrix* — unknown meaning
9. *Canis mustelinus* — Pug

III. **Classification by Stonehenge in British Rural Sports (1856)**
1. Dogs that find game for man, man kills the animal. Example: Spaniel.
2. Dogs that kill game when man has found it for them. Example: Greyhound.
3. Dogs that find game themselves and kill it themselves. Example: Foxhound.
4. Dogs that retrieve game that has been injured by man. Example: Retriever.
5. Useful companions of humans, guard dogs and watch dogs. Example: Mastiff.
6. Lapdogs for ladies. Example: King Charles Spaniel.

We find additional classifications of the breeds with Buffon (1798), who proceeded from the false assumption that the origin of all dog breeds can be found in the sheepdog. The division of his groups proceeds from particular physical traits, such as the length of the legs, ear position, head shape, and the like. Studer (1901), on the basis of the skull forms and basilar lengths of the various dog breeds in comparison to finds of ancient skulls, arrives at a completely new classification, visually oriented, which had numerous supporters for many years. In recent years, however, his theories have been attacked by renowned scientists on grounds that modern skull collections of the same dog breeds display enormous variation, from which earlier scholars using Studer's system would without second thought have inferred the existence of different groups of breeds. Accordingly, the usefulness of the dog skull as a character for classifying groups of breeds has been called in question.

We can only touch on the subject of the scientific classification of breeds within the scope of this book, and not treat it in depth. My main purpose was to use the various classifications to show that many breeds already existed centuries ago.

All scholars certainly concur that the breeding of dog breeds represents a unique cultural achievement of human beings. It demonstrates intellectual open-mindedness, the striving for improvement, and rational planning. The criterion for selection in the origin of breeds was the performance of the dog. Dogs were bred for speed, alertness, stamina, scenting ability, health, and other performance factors. Selective breeding, corresponding to the requirements for performance, slowly yielded the breed types, both in performance and in anatomy. With the development of the breeds of domestic dogs came division of labor among our dogs.

Now we will turn from general observations to our specialists, the fighting dogs. They are the focus of this book.

2. THE ORIGIN OF THE FIGHTING DOG BREEDS

The ancestors of the large breeds of

fighting dog were, with some degree of certainty, the large races of wolves. There are finds of wolf skulls in central Russia and Poland with basilar lengths of 238-244 millimeters. The skulls of modern large dog breeds exhibit similar sizes. Keeping in mind the reservations we had about using skull measurements as the sole differentiating characters of breeds, the existence of similar basilar lengths in races of wolves and large dog breeds proves that, on the basis of size alone, wolf races existed in the past and still exist today that could be the foundation forms of our large fighting dog breeds.

We disagree with the theories of a number of dog researchers that we can localize the origin of the large dog breeds in Tibet and Nepal, and later in Egypt. According to these theories, these dog breeds then spread slowly to many other regions as the result of human migration and trade. This scenario does not seem likely to me. It certainly seems correct that the large-framed, black Tibetan Wolf is the foundation breed of the Tibetan dog. In my opinion, however, certain scholars infer too much from this and the fact that the Chinese literature (Chouking) reports as early as 1121 B.C. of a Tibetan dog that had been tamed by people. They concluded, on the basis of later historical finds, wall murals, ceramic jars, skull finds, and so forth, and from descriptions in the literature, that this sole foundation breed later spread slowly to other regions. In my opinion, however, all we can conclude from this is that the cultural development in the regions in which the first accounts exist was in part more advanced, which is why we have the pictorial and literary portrayals. At the time, or even earlier, large breeds of fighting dog certainly could also have existed in culturally less developed regions. The difference was that they were not mentioned or depicted.

Studer and Strebel's theory that the large dog breeds are the products of large, primitive dog stock, and that they could have developed independently of one another in various places, is largely unprovable. According to this theory, these dog breeds originated particularly in those regions where the human living conditions were characterized by the struggle against large, wild animals and the threat from hostile neighbors.

We must also take into account the theories of other scholars, such as Dr. Keller and Prof. Kramer. It is recognized that through large-scale human migrations and the slow development of trade, and last but not least through gifts between royal courts, valuable fighting dogs, in particular, found their way into new regions and complemented the local stocks. Strong and brave dogs were valuable gifts in this developmental phase of humankind. Therefore, we repeatedly find accounts of military campaigns with fighting dogs and royal gifts in the form of large dogs. Thus, precisely with the fighting dogs, rather early breeders began exchanging these valuable dogs. We should not interpret this to mean, for example, that purebred Tibetan dogs were now bred separately in other regions. Instead, the dogs that were imported, received as gifts, or taken as booty merely served to complement the local large dog breeds.

At this point I would like to make a second, rather important observation. By fighting dogs we must not consider only the giants among dog breeds, but

II. INTRODUCTION

English fighting dog. Bronze from the early nineteenth century. Photo by Lazi Pereny. Personal collection of Dr. Fleig.

rather all breeds with a character suitable for protecting humans and for fighting wild animals. We find here a wide selection of dogge-like giants, approximately in the form of the English Mastiff, down to the proud miniatures for protecting the housewife, such as the Pug. In animal fights, in particular, which will play an important role in this book, we repeatedly encounter the small, fearless fighting dog. We therefore must consider the concept of the fighting dog rather broadly, although the giants, with their enormous body sizes, naturally first draw our attention.

The foundation breed of the fighting dog in its outward appearance naturally was a large, low-slung, heavy breed with a very powerful build, strongly developed head, and tremendously threatening voice. The outward appearance alone of this dog must have instilled fear in its enemies. It was a picture of elemental force, not particularly quick in its movements because of its weight, but powerful in its initial charge. Studer describes the breeding of large fighting dogs as follows: *Even in antiquity, from the dog man attempted to produce breeds, which through their strength and their powerful biting apparatus were able to defend him or his herd against strong enemies, to overpower and pull down large animals on the hunt, and to control large, unmanageable domestic animals. To produce such animals, large dog forms were bred primarily for the development*

Head of an English fighting dog. Bronze from the early nineteenth century. Photo by Lazi Perenyi. Personal collection of Dr. Fleig.

of their biting apparatus. The jaws could develop more power the closer the end point of the fulcrum of the jaw was to the point where the power was applied; that is, the shorter the jaw was in front of the point of attachment of the jaw muscles... The shortened jaws became thicker and heavier with the strong development of the teeth. These circumstances demanded the significant development of the jaw musculature, and this, in turn, a larger point of attachment . . . The shortened snout often caused the nose to be deeply set, making the muzzle appear upward raised. The skin of the face normally did not contract the same amount as the bony foundation shortened, so that facial folds, low-hanging flews, and folds at the corners of the eyes formed, thus giving the entire physiognomy of the dog a rather frightening appearance.

In his classification of the various breeds of fighting dog, Strebel (1905) comes up with five foundation breeds from which an additional eight breeds were developed. This results in the following picture:

I. Tibetan Dog

II. Mastiff
1. Bordeaux Dog, from it the French Bulldog.
2. English Bulldog, from it the Miniature Bulldog.
3. Pug.

III. Great Dane
1. Danish Dog
2. Boxer

IV. Newfoundland

V. Saint Bernard, from it the Leonberg Dog (from IV + V)

With this classification, I must point out that at the turn of the century schematic divisions of this kind were still influenced greatly by nationalistic questions, such as with the assumption that the Great Dane is the ancestor of the Danish Dog. The French Bulldog is certainly much more closely related to the Miniature Bulldog than to the large-framed Bordeaux Dog. The Pug stands much closer to the Bulldog than to the Mastiff, and, in my opinion, the relationship of the Newfoundland and Leonberg with the fighting dogs cannot objectively be maintained. Furthermore, Strebel misunderstood the fundamental position of the Bull and Terrier in Britain. Although it is a cross, it became the mighty root for extraordinarily important new breeds of fighting dogs.

Therefore, I have consciously undertaken my own classification and evaluation of the old fighting dog

II. INTRODUCTION

breeds. They are the foundation forms of numerous modern dog breeds.

The following classification of the original fighting dog breeds seems sensible and useful to me. These breeds largely still exist today. Only one, the Chincha Bulldog, has died out. The others are the foundation forms of the modern fighting dog breeds. I subscribe to the following classification of the original breeds of fighting dog:

1. Tibetan Dog
2. Molossus
3. Bull Biter
4. Great Dane
5. Mastiff
6. Bulldog
7. Bull and Terrier
8. Chincha Bulldog

III. THE PURPOSES OF FIGHTING DOGS

1. WAR DOGS

Whoever wants to write the history of the war dog must write the history of human development from the beginning, since the dog has always been man's closest companion. This is how H. Lloyd views the task of writing this chapter.

I have previously described how man, threatened by hostile neighbors, recognized very early on the value of the dog as an ally against all attackers and as a willing helper on military campaigns and when hunting game. As long as there were wars — until the invention of firearms — we find fighting dogs as a weapon of the attacking warrior and as a defensive shield of those under attack. The brave dog, strong and aggressive, was a highly prized weapon, which often decided victory or defeat. The task of the war dog was to attack all enemies: to knock them down, to neutralize them, even to kill them. The dog — like the master — often was equipped with metal plates and chains. Collars with sharp spikes or hooked knives on the outside protected the dogs and pierced the legs and bodies of their foes. Particularly against enemy riders and their horses, these curved knives inflicted serious injuries.

History offers a wealth of reports on fighting dogs and how they were used. I will mention only a few examples here. Hammurabi, the King of Babylon (circa 2100 B.C.), equipped his warriors with huge dogs. These fighting dogs aroused fear and terror among the enemy. The Lydians, an Asiatic tribe, deployed a separate battalion of fighting dogs in their wars against the Kimmerians (628 to 571 B.C.). The heavy war dogs used by King Alyattes slaughtered the enemy dogs and warriors.

The Persian King Kambyses used huge fighting dogs in his conquest of Egypt. The pack of dogs attacked the Egyptian spearmen and archers (525 B.C.). The battle of Marathon (490 B.C.) is famous. Here the dog of an Athenian so distinguished itself that both were honored as heroes and immortalized in a mural. In the siege

Three sketches of armored war dogs.

III. THE PURPOSES OF FIGHTING DOGS

Armored war dog.

of Mantineia (385 B.C.) by the Spartans, they gained victory only after they completely cut off the reinforcements from the besieged city through the relentless use of their fighting dogs. Interestingly, we find numerous dog bones in Athenian war graves. On his campaign in Greece, Xerxes brought along dogs, which determined the outcome of many battles. At the battle of Vercella (101 A.D.), the Romans under Consul Varius encountered large Kimber dogs, led by women, which led a bloody defense of their laagers. This prompted the Romans to employ fighting dogs of their own in their legions, one dog company per legion.

When the Romans landed in Britain, the giant, wide-mouthed dogs of Britain protected the women and children of the natives. These dogs proved far superior to the Roman's own dogs and many of them were later brought back to Rome as booty for exhibition fights in the Colosseum.

In the great war for France, King Henry VIII reinforced King Carl V of Spain with an auxiliary army of 400 soldiers accompanied by 400 powerful Mastiffs equipped with wide war collars. These animals became such important allies for the Spaniards that Carl V commended these dogs several times and praised them to his soldiers as an ideal of unconditional bravery.

III. THE PURPOSES OF FIGHTING DOGS

Old Assyrian dogge (from Lazard).

In the siege of Valencia there was a bloody battle between the defending French and the attacking Spanish-English dogs, which ended in the defeat of the besieged French.

Just before the battle of Aboukir in July 1799, Napoleon ordered his general Marmont to assemble a large number of fighting dogs and to array them in front of his reserves. In Britain, during the reign of Elizabeth I, 800 fighting dogs helped the Earl of Essex to put down the Irish uprising in bloody fashion.

Fighting dogs have also been used, however, in battles in other parts of the world. The Spaniards, in particular, used their dogs during their conquest of the Americas. During the fifteenth century, soon after the discovery of the New World, powerful fighting dogs were used to inspire fear and terror among the native Indians. Peter Las Casas reports of a battle at La Vega Real where the dogs vanquished the natives, with each dog alone incapacitating more than a hundred Indians. These dogs had been made blood-thirsty for human blood and turned the countryside almost into an abandoned wasteland.

I offer these documented historical uses of fighting dogs only by way of illustration. Certainly they strengthen the hypothesis that in the innumerable battles and wars that characterize our human existence, fighting dogs have often been used to attack human beings.

From the historian Aldrovandus (1637), who in turn draws on the Italian historian Blondus (1388-1463), we learn about the breeding and training of these dogs: *The war dog must have a frightening appearance and look as if it is always ready to fight.*

Assyrian dogge: Flat relief in the Palace of Niniveh.

Everyone — with the exception of its master — is its enemy. This also means that it will not allow itself to be touched, even by someone it knows well, but will threaten to sink its teeth into anyone's body. It must stare at everyone with a hostile look and appear to all people as if filled with inner rage. This dog is trained to fight from its earliest youth. For this purpose a person is equipped with a cloak of thick pelts, which the dog cannot bite through, and the man protected in this

III. THE PURPOSES OF FIGHTING DOGS

Molossus on the altar relief of Pergamon.

way becomes the practice goal of the dog. The dog is roused and the man runs away. He is caught, pulled down by the dog, and mauled fiercely while lying on the ground. The following day the dog is set on another man, who is also protected, and at the end of the training the dog can be set on any man.

After the fight the dog is chained and is subsequently fed while chained. These exercises are continued until the dog has become a first-class defender of humans. Blondus is even of the opinion that from time to time it is proper to fight the dog with drawn sword. In this manner the dog supposedly develops extreme understanding and courage. Subsequently, we can set the dog on any foe. It is downright amazing how a substantial part of the training for guard dogs still practiced today was already used systematically in the training of ancient fighting dogs 600 years ago!

We cannot leave this chapter without a brief aside to the war dogs that have completely different tasks than the direct attack of the enemy. The Mederians and the Persians deployed the dog as a messenger through enemy lines. Here the dog was used ruthlessly as a means to an end, in that the messenger dog swallowed the message and was slaughtered by the allies upon its arrival. Later, starting about 600 B.C., the message was fastened to the collar instead of being swallowed. There was now a risk that the message would fall into enemy hands, but the life of the faithful helper was spared.

In recent wars, dogs have been given new duties as medical dogs, munitions dogs, mine-detecting dogs, and guard dogs. In World War I, Germany used about 30,000 war dogs and France about 20,000. Dogs were also used in many other countries. In World War II, in Russia alone, more than 50,000 working dogs served in the military. In the Algerian War in 1953, about 7500 dogs were still in service. Our ancient fighting dog breeds, however, were no longer used for these tasks; rather, the modern

Canis bellicosus from Gessner.

III. THE PURPOSES OF FIGHTING DOGS

War dogs in the American Civil War (illustration from an American magazine, 1867).

III. THE PURPOSES OF FIGHTING DOGS

working dog breeds were recruited. The fighting dog breeds were hardly suitable for these new tasks, because their strong fighting and protecting instincts stood in the way of their proper training. In this way they escaped the bloodshed to which the other breeds were exposed.

The heyday of the war dog is over. The fighting dog of ancient pedigree lost its military purpose well before the Atomic Age, in fact with the introduction of modern firearms. Anyone who has feeling for heart and courage must admire the dog that will fight to the death. It does not shoulder the moral blame for all the bloody wars in which it took part. It was bred and trained for the reckless defense of the master and was man's tool. Man alone must take the blame for misusing its protection in many places. The fighting dogs served man faithfully as war dogs for millennia.

2. THE HUNTING OF DANGEROUS GAME

The hunting of dangerous game has always been a challenge for man. Over many centuries, the aurochs, stag, wild boar, and bear were the focus of great hunts. The feudal hunting right reserved these hunts largely for kings, the nobility, and large land owners. Princely courts had impressive hunting packs, among them large hunting dogs for hunting dangerous game and fast tracking dogs.

The European forests harbored two kinds of large cattle, the aurochs and the bison. Up to the beginning of the Middle Ages, the aurochs still ranged over all of Central Europe, but then retreated more and more to the East. Up to about 1630 we find accounts of aurochs hunts. After that this ancestral form of cattle appears to have become extinct.

The bison, as a true forest animal,

Hunting scene: Antonie Tempesta (1555-1630).

III. THE PURPOSES OF FIGHTING DOGS

lived on in European forests long after that. As late as 1860, Czar Alexander II killed 28 bison on a great hunt. In 1921 the last reported hunt took place in the wild in Poland. Both species are bred today in zoos, and attempts are being made to breed these ancient animals in their old form.

The hunting of aurochs and bison was a royal right. The men opposed these dangerous foes with spear in hand, protected by a large tree. The dogs' job was to position the game so that the hunter could kill it with the spear. Of Charlemagne it is reported that he especially loved this dangerous hunt. The Meissen porcelain figurine from the masterly hand of Johann Joachim Kaendler (1706-1775) shows the power and beauty of this game and the powerful hunting dogs, adorned with gold-decorated collars, a sure sign of the high standing of their owner.

Swift stags, followed by long-legged tracking dogs, behind the pack the powerful hunting dogs to seize the game, then the hunter on horseback, this was the lordly *high hunt* of the noble game, the stag. The engraving by Johann Elias Ridinger, from about the year 1729, gives us an impression of the end of this hunt and of the dogs that killed the stag. It is not difficult to imagine how dangerous this hunt was precisely for the hunting pack. The antlers of the threatened stag became a lethal weapon against the dogs.

Our fighting dogs were mainly used to hunt the wild boar. Ridinger writes about the boar pack: *This is the most comical but also the most dangerous of hunts, because a wild pig is such a dangerous animal that it spares neither man, horse, nor dog. When it is cornered in the thicket, sheets and cloths are raised on the sides so that it can be forced to the best place to run. Then the boar finder or boar dog is let loose and the boar tries to break out. Now the hunters shout to the dog, 'Look out, look out!' and others call out, 'Set to, set to!' Now the light dogs are let loose, which spin the wild boar around and tire it out. Then the heavy or English large dogs are let loose and hold it until the hunters can dispatch it with the hunting knife...* Ridinger's engraving *The Boar Hunt* (1729) illustrates this text. On the same subject, I also show *The Boar Hunt* from the *Jacht Buch Wolff Pirckner von Bayreuth* (1639).

Johann Tantzer, in the book he wrote in 1699, *Der Dianen hohe und niedere Jagtgeheimnisse*, gives quite useful insight into dog keeping in the royal courts. He reports of heavy, large English Mastiffs. These were divided into chamber and ordinary dogs. The chamber dogs were the best of their kind. As a sign of their rank they carried silver-plated or gilded collars adorned with fringes. On the hunt they were protected by armored jackets (on this subject see Ridinger's engraving *Gepanzerte oder mit einter Jacke bekleidete Leib- und Cammerhunde*). Tantzer describes this armor: *The outside is made of brown cloth or silk, lined with strong linen, stuffed full with hair or cotton, and sewn together. It is not stuffed under the belly and chest, however, because it is the most dangerous there. Rather, it is equipped with whalebone and sewn together tightly with much effort, so that it is as strong as armor.*

A look back in history shows our fighting dogs in the second century A.D. Oppian reports in his book *De Venatione* about various dog breeds and writes: *Other dogs in turn are wild and resist the force of any attacker,*

III. THE PURPOSES OF FIGHTING DOGS

An aurochs hunt: from J. H. Ridinger, lithograph H. Menzler.

III. THE PURPOSES OF FIGHTING DOGS

Aurochs hunt: Antonie Tempesta (1555-1630).

particularly in the hunt of the aurochs, wild boar, and lion. These dogs are flat faced and have fear-inspiring wrinkles, which droop above the eyebrows. Below them there are flashing eyes, which smolder like fire . . . Of all hunting dogs they are the best breed. In their outward appearance they resemble closely other meat-eating predators, such as wolves, tigers, or leopards . . . Like these, they are fast and very strong. We see that over many centuries, dogs of similar form and the same breed stood by man's side in many countries of the world. They are also shown in a detail from a Pompeian mural and hunting scenes from the tomb of Scaurus, as well as in a detail from a Spanish ceiling mural of the Alhambra (fourteenth to fifteenth century). In Germanic law from the seventh century, we find strict penalties to protect *canis porcatorius* (boar dog).

The great hunts tell of the performance of the hunting dog of the Middle Ages. In 1556, Count Philipp in the Habichts, Reinhards, and Kaufunger forests, had killed 726 wild boars with his dogs by November 30. In 1559 the number was 1120 and the year 1563 brought a record total of 2572 wild boar, including 253 boars, 1145 sows, and 1174 young. On a single hunt by Duke Karl von Wurttemberg in the year 1782, 2600 wild boar were killed.

This hunt was extremely dangerous for man and dog. Catching the wild boar with the boar dogs required not only courage but skill and luck as well. Many a hunter paid for this hunt with his health, even with his life. The

III. THE PURPOSES OF FIGHTING DOGS

loss of life among the dogs was enormous. *If you want pigs' heads, you must use dogs' heads!* This adage is illustrated by Snyder's *A Raging Boar* and by Jakobsen's *The Wild-Boar Hunt.*

The largest number of hunting dogs is attributed to Duke Heinrich Julius von Braunschweig, who in 1592 hunted with 6000 boar dogs on the upper Weser River. At virtually all castles of the nobility, we find separate dog houses for the large pack of dogs. Because the dogs kept by the nobility themselves usually did not suffice because of the high losses among the dogs, by official decree the vassals were obligated to supply boar dogs to the court. Shepherd, miller, and butcher had to breed dogs for their lord. Shepherds who delivered unfit dogs to Count Moritz the Erudite, for example, had to pay a penalty of five lambs. Because of the great damage caused by game from an agricultural point of view in the nineteenth century, the population of wild boar was reduced and the hunting packs became smaller and smaller. Furthermore, the further development of firearms decimated the wild boar and largely made the boar dogs superfluous.

That the manly fight with the pack of dogs and the boar knife has not died out even today is shown by reports of hunting with Bull Terriers in the great forests of the Czech Republic and Germany. The small and athletic Bull Terrier has proved to be particularly suitable for this hunt. It seizes the wild boar by the head and

Exhibition fight of aurochs versus fighting dogs and gladiators, engraving by Stradanus (1523-1605).

III. THE PURPOSES OF FIGHTING DOGS

Johann Elias Ridinger: *Hunting a stag with hounds*.

III. THE PURPOSES OF FIGHTING DOGS

III. THE PURPOSES OF FIGHTING DOGS

Johann Elias Ridinger (1698-1769): Prince's ordinary and chamber dogs, armored or with a vest.

holds on mercilessly. At a weight of 20-30 kilograms, this dog is also able to slow down the wild run of the wild boar. In so doing it gives the hunter the opportunity to stab the wild boar with the knife. So, near the end of the twentieth century, we still find the huntsman of the Middle Ages!

In closing, I must mention one of the first documents of the boar hunt that is still preserved today. This is the oil painting by Abraham Hondius. Here we see from the year 1685 two Mastiffs attacking the fleeing boar in northern Scotland. With this marvelous picture we conclude the discussion of the gallant hunt of the wild boar by our brave dogs.

Now let us turn to bears. Originally the brown bear was the largest and most dangerous predator throughout Europe. As an omnivore, it makes do with a diet consisting predominantly of carrion and vegetarian food. Although it occasionally took sickly game or set upon man's herds, the damage it caused to man was minimal. For the hunter, hunting the bear was a noble game. At his side we find the large Bear Biter. Pairs of these powerful and aggressive dogs were usually leashed in teams. Their task was to corner the bear and to distract it through its attacks so that the hunter could kill the bear with the lance or hunting knife. This hunt is shown in a marvelous engraving by Justus Sadler, *On the Bear Hunt*, from about the year 1650. The large hunting dogs, short-faced, heavy-boned, and full of aggression, offer an

A raging boar: Frans Synders.

III. THE PURPOSES OF FIGHTING DOGS

Fighting scene of a boar dog versus a boar. Bronze from the early nineteenth century.

impressive picture of the fighting dogs of this time.

On these hunts, which took place in inaccessible swamps, forests, or mountains, the bear was usually cornered. The hunting party consisted of about five teams of two Bear Biters each. First one team was used to drive the bear to the hunter or to keep it at bay. If a bear showed itself, more dogs were let loose, and the hunter killed the cornered bear with the hunting knife. In the Middle Ages the bear hunt was considered a heroic deed. To be armed only with a lance and dagger was the Royal Game for the nobility. We have a particularly impressive artistic portrayal of the bear hunt by Samuel Howitt (approximately 1800). It shows us the large-framed bear dogs

III. THE PURPOSES OF FIGHTING DOGS

A boar hunt: by J. E. Ridinger, lithograph H. Menzler.

III. THE PURPOSES OF FIGHTING DOGS

Boar hunt in the English Colonies: engraving, 1863.

Juriaen Jacobsen: A wild pig attacked by dogs.

III. THE PURPOSES OF FIGHTING DOGS

Ludwig Beckmann *Boar hunt* (woodcut).

III. THE PURPOSES OF FIGHTING DOGS

III. THE PURPOSES OF FIGHTING DOGS

Detail from a Pompeian mural.

and the danger of the attacked bear. Franz Snyder's *Bear Hunt* and *Bear Fight* also depict similar scenes. Specht's *Bear Hunt* also gives a superb impression of this regal hunt.

Before we turn to several specific descriptions of the hunt, here is a summary of the characteristics of the dangerous game that our fighting dogs contended with:

Aurochs and bison: Height at the withers about 180 centimeters, length 350 centimeters, weight 500-700 kilograms. Weapons: horns and hooves. High degree of intelligence, quick to defend itself, always ready to attack.

Stag: Shoulder height 120-150 centimeters, length 180-220 centimeters, weight 160-270 kilograms, antlers as dangerous weapons at lengths of 86-120 centimeters. Attempts to flee. A tenacious, very fast runner. When cornered, prepared to defend itself with antlers and hooves.

Wild boar: Shoulder height up to 102 centimeters, length about 150 centimeters, weight 150-200 kilograms. Fights with tusks (weapons), body weight, high degree of intelligence, great speed and maneuverability. Always ready to attack.

Bear: Height at the withers about 100-125 centimeters, length 200-220 centimeters, weight 150-250 kilograms, large specimens up to 350 kilograms. Dangerous teeth, frightful paw blows. Highly intelligent.

Our bear dogs, boar dogs, and bull biters seemingly have little with which to oppose these animals, which are far superior to them in size and weight. What they do have, however, is decisive: strength, insensitivity to pain, stamina, and a tremendous will to win. Therefore, together with their master, they become the most dangerous enemy of this dangerous game. Admittedly, innumerable dogs lost their lives on these hunts, but the survivors were the winners. They were the dogs that contributed to this

Hunting scene from the tomb of Scaurus.

III. THE PURPOSES OF FIGHTING DOGS

Ceiling mural in the Alhambra.

excellent dog lineage in breeding. From this breeding material, the product of strict selection in the hunt for dangerous game, the large fighting dog breeds were created.

It appears extraordinarily interesting to us, to complement the descriptions of the hunt in Europe, to provide a report from the Indian subcontinent. In the book *Thirteen Years Among the Wild Beasts of India*, the Englishman George P. Sanderson reports on the *Dangerous Game Hunting with Dogs* during the 1870s. Here English fighting dogs, all from the Bull and Terrier breed, were used to hunt bears, buffalo, elephants and panthers. The author stresses that this hybrid breed in particular was superbly suited for these hunts because of its character. This breed combines superbly the Bulldog's unconditional courage and its direct attack with the speed and intelligence of the terrier. I will discuss these points individually in the descriptions of the breeds. Our big-game hunter used six Bull and Terriers in his pack, each weighing about 15-20 kilograms. Let us accompany him on his hunts through India.

One morning my men managed to spot two bears in a favorable location on a hill among some rocks . . . We approached the rocks from the side . . . When the bears saw us at a distance of about 30 paces, they tried to escape. I killed the female bear with a single shot. She fell off the rocks and rolled a short distance. The bear jumped over her and ran roaring down the hill. The dogs were still leashed. As soon as the bear took flight, they were loosed. Unfortunately, Turk fell upon the dead female, while the other five dogs set off after the escaping bear. Marquis was the first to reach the bear and jumped at its head. The two

Armored hunting dog, from Tantzer.

III. THE PURPOSES OF FIGHTING DOGS

Ludwig Beckmann: *Dispatching*. Woodcut.

A bear fight: Frans Snyders.

III. THE PURPOSES OF FIGHTING DOGS

The Bear at Bay: Samuel Howitt, 1803.

tumbled over, the bear roaring loudly. I did not think any dog could hold on during such somersaults, but when both came to a stop, Marquis still clung tightly to the bear's head. The bear now raised up on its hind legs and would have knocked the dog off with its paw, but at that moment Bismarck seized one ear, Lady the other, and Viper and Fury seized it by the snout. From this moment on the bear was no longer able to defend itself, the dogs pulled it over sideways, and pressed its head against the ground. It tried mightily, but in vain, to free its head with its paws. The dogs continued to press its head against the ground, so that it could not swing its paws far, whereby it could scarcely injure the dogs with the swipes of its paws. Within ten minutes the bear was exhausted, and two thrusts of the knife behind the shoulder finally killed it . . None of the dogs suffered as much as a scratch in this fight.

Our big-game hunter also used his pack to hunt wild buffalo in India. On the basis of his experiences, he was certain that a bison or wild buffalo had no chance at all against three or five fighting dogs. Its enormous strength was of no help against such opponents.

On August 30, 1876, he set out to

Fr. Specht: *Master Brown in distress* (woodcut).

III. THE PURPOSES OF FIGHTING DOGS

H. J. Ueka: *Bear hunting with Dogs*.

Frans Snyders: The bear hunt.

III. THE PURPOSES OF FIGHTING DOGS

On the bear hunt: Justus Sadler, circa 1650.

Bear hunt: Antonie Tempesta, 1608.

III. THE PURPOSES OF FIGHTING DOGS

Bear Hunting: Samuel Howitt, 1797.

hunt with five Bull and Terriers and a number of tracking dogs (finders). After two hours of searching they approached the buffalo, and they released the fast tracking dogs. The Bull and Terriers stayed on the leash until the tracking dogs had cornered the game. They had pursued a single bull until they held it at bay on a hill. Then the Bull and Terriers were released. *I had just spotted the prey when Bill jumped at the bull's nose and held on. Then the bull turned and thundered down the hill. It ran by me at an astounding speed for such a heavy animal. Bill of course held on tirelessly, although he was carried through two or three dense thickets, where any other less determined dog would have been thrown. Bill was the model of a Bull and Terrier, he always went straight through and never waited long for a better opportunity or help from another dog. Occasionally, dogs are killed because of such reckless courage, but in general they are the safest when they break through without hesitation and hold on tight, despite all resistance. How often are curs killed, which have just enough courage to attack when they are within reach of a truly dangerous foe, whereas the dogs that hold on recklessly get away comparatively uninjured with such beasts . . . The other dogs did not get their chance until the buffalo reached the foot of the hill. They jumped at its head, one after another, and pulled it down. Then they held its nose to the ground, while it roared in*

III. THE PURPOSES OF FIGHTING DOGS

Johann Elias Ridinger (1698-1769): Bear hunt in winter.

III. THE PURPOSES OF FIGHTING DOGS

fright and pain. It made several futile attempts to gore the dogs with its horns. Thus, a few dogs proved to be a much more dangerous foe for the buffalo than, for example, a tiger. Because of its size, the tiger can be more easily gored and trampled by the buffalo. The buffalo was killed by the hunter's bullet.

We conclude these reports from India with an elephant hunt. The hunting party was on a bear hunt in the jungle, when they came upon a young elephant, about two years old, which had become separated from its herd. He decided to try out his dogs here too. *The elephant saw us and trotted off, but had not even gone two hundred steps when Lady ran it down and jumped at its cheek, seized it, and held on. Seconds later, Bill and Turk had it by the trunk. It stumbled as Bismarck grabbed it by the top of the ear. As the elephant stood up again and ran away, Bismarck rode on its head, the ear firmly in its teeth, with its hind legs hanging down on the opposite side. The elephant trumpeted loudly and dragged Bill and Turk, which were still hanging on the trunk, along the ground. They often lay on their back and were always in danger of being trampled to death. The jungle was full of dry bamboo, through which the hunters and hunted moved with a tremendous din. All of the small dogs, virtually mad with excitement, barked and snapped at the elephant's hind legs as if they were after a sheep, despite the kicks they received. It appeared quite certain that several dogs would be killed. Bill and Turk had the most dangerous places, but they would let go only if they were killed on the spot.* The hunter had a tame elephant on a rope, about a mile away. This was fetched, while the fight between the dogs and the elephant continued. When the rope finally arrived, Bismarck, Lady, and Bill were still at their posts. Turk, however, was so exhausted that he had been shaken off . . . Had the rope not arrived at that moment, the dogs doubtlessly would slowly have become exhausted, and the elephant would have escaped. By means of two fetters on the hind legs, tied together with a large tree, the elephant was subdued. *When both legs were secured, we had to separate the dogs from the elephant, a very dangerous task, because we first had to catch them, hold them, and slowly pull them off, and all this within the dangerous reach of the elephant. The elephant, two years old, weighed about 450 kilograms, a good catch for dogs weighing about 20 kilograms!*

Our big-game hunter reports of other occasions on which his dogs were used against leopards, bears and panthers. These hunts presented two problems. The first was that there was always the danger that the dogs in their great excitement would fight one another. The other problem was that the dogs could become so exhausted in the long fights against such dangerous opponents, that in the tremendous heat in this country they could become virtually helpless and break down half asphyxiated after the victory. They revived only after being splashed repeatedly with water.

Now a few comments from the typical English point of view on the question of the cruelty of this hunt to the poor dog. *Many people speak of cruelty and pain completely in the abstract without knowing the precise circumstances of the individual case. Some of them would think nothing of keeping a Bull and Terrier, for example, chained for months, but they complain vociferously when such a dog is exposed to the danger of being killed by a wild*

III. THE PURPOSES OF FIGHTING DOGS

Hunting the Porcupine: Engraving by A. Hondious, published 1803.

animal. Anyone who has any idea about such things knows that these dogs in their full excitement feel no pain at all, whereas, when robbed of their freedom and movement, they suffer tremendously. Their natural instincts find their greatest fulfillment in the all-out fight. Could anyone doubt at all which of the two lives the dog itself would choose: the life fighting the wild beast or the life chained in the kennel? Anyone who gives this question serious consideration does not understand fighting dogs at all.

I have presented these detailed reports because they do an excellent job of describing the use of fighting dogs against dangerous game. It should not be difficult for our reader to visualize the hunt in our European forests for the aurochs, stag, wild boar and bear, and to imagine fighting scenes similar to those depicted so graphically from India. Our hunting dog, like the fearless Bull and Terriers, was a willing and always available helper for man in the great hunt for dangerous game. It contributed decisively to man's mastery of these powerful animals and subjugation of the animal kingdom.

3. FIGHTS AGAINST BEARS AND LIONS

In the Greek legends we find the representation of the victory of his dog over two lions on Achilles's shield. The legend also claims — doubtless in support of this image — that the Persian King Kambyses (reigned 529-522 B.C.) possessed a dog that started a fight with two full-grown lions. We have examples here of the eternal dream of the longing for the invincible dog as a faithful companion of man, which defeats even the King of Beasts — the lion.

On the same subject, here is a third story: *Alexander the Great, the King of*

III. THE PURPOSES OF FIGHTING DOGS

Macedonia from 356-323 B.C., reached India on his great military campaign. Aelian, a Roman historian, reports (about 220 A.D.) that the Indians showed Alexander powerful dogs, offspring of tigers, which because of their noble blood refused to fight the stag, boar or bear. Only the fight against lions was appropriate to their rank. First the Indians released a stag, but the dog paid it no attention, then a wild boar, but the dog remained quiet and uninterested. Even a bear failed to change its behavior. As soon as it spied a lion, however, it was filled with grim rage and without the slightest hesitation the dog fell upon the lion, seized it in a powerful hold with its jaws and began to choke it. Then the Indian King Sopeithes, who was well aware of the dog's toughness, commanded that the dog's tail be cut off. The dog did not appear to notice this in the least. The Indian commanded a foot to be chopped off, and this too was done, and again the dog showed no reaction. A second leg was chopped off, but its grip lost none of its fierceness. The same was true after the loss of the third — and the fourth — leg. Finally the head was separated from the rest of the body with a sword, but the dog's teeth held tightly with the same tenacity as at the beginning of the fight.

This account by Aelian doubtless holds the spiritual authorship for many subsequently documented reports of similar events. The reader should be aware that even historians occasionally take up fables in their accounts, which, although they hold

Charley's Theatre Westminster: Bear fight of the old school, Henry Alken, London 1821.

III. THE PURPOSES OF FIGHTING DOGS

certain basic truths, have been so embellished by imagination that in the process certain exaggerations are incorporated. It is true that precisely our fighting dogs are characterized by nearly total insensivity to pain during a fight. During the fight, particularly when they have suffered painful injuries, they go into a kind of *red phase* — a biting frenzy — during which they become completely insensitive to all physical pain. Such a *red phase* is also falsely attributed to the fighting bull, which blindly attacks the red cape. What is untrue of the bull, however, is true of the fighting dog. When fighting they enter a phase in which rage and aggressiveness switch off the sensitivity to pain. Precisely here lies the great danger of these dogs, and causes certain problems in their rearing and keeping. In the course of our further inquiries it will become clear how much Aelian's story has inspired the imagination and actions of others.

Stradanus (1523-1605), in an engraving from Galle's workshop (1578) based on his original mural, offers a superb illustration of the events in the combat arena under the rule of Alexander the Great. Here, before the chosen public in the arena, the mighty fighting dog fights the lion, after having defeated an elephant.

Now, however, we turn to Britain, the center of animal fights in the Middle Ages. For the English nobility, who knew and loved the hunt of the bear in the wild, it was a courtly pleasure to match a bear caught in a trap against their mastiffs. *Bear baiting* became a *courtly sport* in Britain, and we can trace it back as far as the year 1050 on the basis of historical documents. Here Edward the Confessor permitted a fight between a bear and six mastiffs in the town.

Lion fight in the arena in the time of Emperor Alexander: Engraving by Stradanus (1523-1605).

III. THE PURPOSES OF FIGHTING DOGS

The Fight between the Lion Wallace and the Dogs Tinker and Ball at Warwick. From Pierce Egan's *Anecdotes*, London 1827.

Lions, however, princely gifts from foreign lands, also provided entertainment at court. On one occasion, James I (1566-1625), the son of Mary Stuart, staged a fight between a lion and three mastiffs. Here is the account of subsequent events: *One of the dogs, which was the first to be sent in the cage, was soon put out of action by the lion, which seized it by the head and neck and dragged it through the cage. A second dog was sent in and met with the same fate. The third, however, who came to its aid, immediately seized the lion by the lower jaw and gripped it securely for a considerable time until, severely injured by the lion's claws, it was forced to loosen its grip. The lion itself was seriously injured in the fight and was not able to continue fighting. With a sudden mighty leap over the dogs, it fled inside its den. Two of the dogs died shortly after the fight from the injuries they had suffered. The last, however, survived this splendid fight and was nursed back to health with great care by the King's son, Prince Henry. Prince Henry declared: 'He had fought the king of the wild animals and should never again have to fight baser creatures!' In this way the dog had gained for itself a safe life at the English Royal court.*

The center of attraction of these bloody fights naturally was bear baiting, because bears were much easier and cheaper to acquire than lions. For this bear baiting, uniform rules of combat soon developed. Here the bear wore an iron collar, to which was attached a chain. The chain in turn was fastened to a thick rope. The chain and rope ran through strong iron rings and rollers on the wall so that the bear — standing upright — awaited its four-legged adversary. By fastening the bear to the rope in this way, the bear keeper could control the bear by lengthening or shortening the rope without himself being exposed to direct danger. The illustration by Henry Alken, *Bear Baiting,* from the year 1820 shows clearly the fighting posture of the bear and the mechanism the bear keeper used to control the bear. The reader will also find two additional illuminating illustrations. From *Sporting Magazine* of the year 1796 I have taken a portrayal called *Bear Baiting*, a picture with much atmosphere and a fight in the open outside the arena, an unusual practice. There is also a second print by Henry Alken from the year 1821, published in *Real Life of London* by Pierce Egan from Charley's Theatre in Westminster. The print shows a quite proper and bourgeois activity. The original sport of the court had become a sport for the commoner.

Now, however, let us return once more to the English Royal court. Elizabeth I (1533-1603) was a great follower of animal fights. She was especially interested in the mastiffs she herself bred. Bear baiting found tremendous support during her reign. From old accounts we know that on May 25, 1559, Elizabeth received the French ambassador at court, enjoyed a superb meal accompanied by courtly tea-time music, and after that was wonderfully entertained by fights between powerful English dogs and bears and bulls.

Her Royal Highness and the ambassador stood together in the gallery and enjoyed the fights until late in the night. The diplomat was so taken by this spectacle that from then on Her Majesty never neglected to stage a similar show for foreign guests she particularly wanted to impress. At the famous festival in Kenilworth, which was arranged in honor of Queen Elizabeth, numerous animal fights took place. During the six days of the

Bear Baiting: print published in *Sporting Magazine*, 1796.

III. THE PURPOSES OF FIGHTING DOGS

Bear Baiting. Now Master George — Let go fair. Henry Alken, London 1823 (detail).

festival, fights between thirteen different bears and numerous mastiffs took place, which the Queen — according to the court accounts — enjoyed immensely. From this time we have the account of an eye witness, who observed a *big fight day* in the summer of 1575, in which thirteen bears fought with dogs. Mr. Laneham writes: *It was a very pleasant, sporting event to witness here how the bear with small, bloodshot eyes awaited the attack of its adversary. The courage and pluck of the attacking dog were astounding, as was the skill of the bear, which usually was able to avoid serious injury. If the bear was really bitten somewhere, it immediately attacked energetically to free itself again. If a dog managed to get a firm hold, the bear cleverly got loose, thanks to its bites and swipes of its paws, accompanied by hissing, roaring, shoving, and squeezing. According to accounts, the bear seldom suffered more serious injuries, which did not heal within a month through careful licking.*

We can easily imagine that even when the bear was sent into combat equipped with a muzzle, it would be able to injure numerous brave dogs so badly with its claws that they died from their wounds. This combat took a heavy toll in blood from the brave fighting dogs. In many cases, however, the bear was also bitten to death by the mastiffs.

Queen Elizabeth's successor, James I, further promoted these animal fights. Under his rule the *Master of the Game Beares, Bulles and Dogges* received an annual income of 450 pounds, an enormous sum at that time. This Master and his men were responsible for all animal fights at the English Royal court. The acquisition, care, and breeding of the animals were his duties. At the King's bidding, at least twenty mastiff bitches had to be kept for breeding in the Tower of London alone as the basis for the royal breeding of *bear dogs*.

In the times of the reigns of Queen Elizabeth and James I, a number of Bear Gardens were built in London. These were arenas for animal fights, which now were also open to commoners. The oldest of these places was probably the Old Bear Garden in Bankside, Southwark, with a first historical reference in 1574. This building is located in the center of London on the southern bank of the Thames and still exists today — to be sure in a new form — as the Bear Garden Museum. This building originally was round and had no roof, apparently in imitation of the ancient Roman amphitheater. In these buildings, bears, bulls, and other wild animals were kept for the animal fights. We find built-in bear dens, occasionally even connected to water holes, so that the bears could bathe. Around the center, where the pit — the arena — was located, rose the stands for the spectators. The animals were housed under the stands.

After Oliver Cromwell (1599-1658) seized power, the Puritans banned the animal fights. After the Restoration, these fights were revived to even greater popularity, and found more

Bear Baiting: Henry Alken, London 1820.

III. THE PURPOSES OF FIGHTING DOGS

Bull Baiting Scenes: Henry Alken, London 1817.

and more followers, particularly among the masses. People were of the opinion that in the duel between bear and mastiff, an ancient hereditary hatred was always carried out anew. Here the bear had the advantage of size, superior strength, and a thick pelt, which made it virtually invulnerable. It was a fight between teeth and claws.

Often two dogs were simultaneously set on the bear. Sometimes the bear received additional protection from a spiked collar on the throat, because the valuable bears were needed for many fights. In other fights, the bear was fitted with a muzzle, so that it could defend itself only with its paws, thus sparing the life of the dog. In this connection it is quite interesting to note that the swipes of the bear's paw are the more deadly, the farther away the adversary is. Once the dog has seized the bear by the throat, because of the loss of leverage, the bear cannot do nearly as much damage to the dog with its paws.

We can only understand the intensity of this *sport* and its popularity among the masses by knowing that the typical English passion for gambling concentrated on the outcome of the fights. Bets were made between the spectators on whether a particular dog could seize the bear by the neck, how long it could hold on, and so forth. For this purpose there were betting sheets with accounts of previous fights and their outcome, both of the bear and the dog. In this connection I must stress that the bears were kept professionally in the Bear Garden. The bear's owner, for a certain minimum stake, allowed dog owners to set their dogs on the bear. These stakes of the dog keepers and the entrance fees of the spectators made an experienced bear a valuable commodity, which could make its owner rich. Individual bears also became legendary for their skill in warding off the dogs. From Shakespeare's time, there is an account of a bear named *Blackface*, which was usually protected by a wide, iron collar in his fights. Equipped with this collar, the bear fought 22 fights against famous mastiffs and defeated them all. Finally, without the protective collar, it was attacked simultaneously by three huge mastiffs, which killed it. Its fame preserved for future generations in an old English ballad.

The contemporary literature proves to be so interesting, so thick with the atmosphere of the fights, that it would be a pity to replace this colorful prose with the author's modern, dispassionate manner of expression. So, for the benefit of my readers, I will embellish this chapter with two contemporary accounts.

In the book *Real Life in London*, written in 1821 by an amateur, we find a first-rate description of a bear baiting. Two visitors arrive at the Westminster Pit. The audience was quite mixed: butchers, dog fanciers, ruffians, gentlemen in modern dress, fops, street vendors, coal carriers, ferrymen, soldiers, and liveried servants. *When we entered, the bear had just been gripped on the head by a dog.*

Wild boar hunt. Oil painting, eighteenth century, unsigned.

III. THE PURPOSES OF FIGHTING DOGS 57

Bear Baiting, hand-colored print, Henry Alken, London 1823.

The dog was owned by an ardent disciple of this sport, who with arms raised high declared to all that this was a damned good grip, really a first-class grip, so help him God. This dog, with its life at stake, had gripped the poor bear, Bruin, by the lower lip. Bruin unleashed a mighty howl, clearly revealing the degree of its suffering. It then tried to give the dog a fraternal swipe of the paw. The many other dogs outside the ring scolded loudly in their desire to take part in this fun. The bear was fastened to a hook on the wall by a chain from the collar, and was forced to stand almost upright. It shook its adversary mightily in ferocity and rage, because of its intense suffering. Now more bets were made and watches were pulled out, to tell precisely how long the 'doggie' could continue to torment the bear.

The dog breeders outdid one another in the appraisal of their prize dogs. Each held his darling between his knees, to accept a favorable opportunity for a bet, and many bets were won or lost in a short time. Bob remained a silent spectator, whereas his cousin, who had a better understanding of the rumor mill, mingled with the poker-faced sportsmen, inquired about the names of the dogs, what victories they had previously achieved, when they had last fought, and other equally important questions for amateurs.

Concerning the further course of the evening at the fights it is reported: *Bruin rid itself of the dogs that subsequently appeared against the bear as fast as they approached. If one of them happened to get a grip, it received a swipe of the paw that almost cost it its life, but at least knocked all the strength*

III. THE PURPOSES OF FIGHTING DOGS

out of it, so that it could no longer hold on. When Bruin was finished with an adversary, this was merely the signal for the next dog to attack . . . When Charley felt that his bear had had enough exercise for the evening, he took it back to its cage, covered with scratches and almost lame, to recover from its wounds, with the single goal of being ready as soon as possible for new fights.

From the year 1825, we have a particularly impressive account of a fight between Wombwell's lions and fighting dogs. Here, through newspaper advertisements, brave dogs were sought to fight against Mr. Wombwell's lions. Twenty-four dogs reported for the fight in Warwick. Mr. Wombwell selected from these, whereby he cleverly avoided the largest, so that the lions would only face medium-sized dogs. According to accounts and illustrations, these dogs were clearly of the Bull and Terrier breed, a new breed at that time. The fights were held in the Old Factory in Warwick, whereby a large steel cage, which was built expressly for this purpose, formed the combat arena. E. S. Montgomery found a quite graphic account of the first fight between Nero and three Bull and Terriers in an old newspaper. This fight took place on July 26, 1825. Here follows the account: *Nero, the lion, was a tame specimen, which had been raised from a cub by humans with loving care. Although Nero stayed on the defensive at*

Bear hunt. Samuel Howitt, oil painting circa 1800.

III. THE PURPOSES OF FIGHTING DOGS

first, he gave his adversaries a fearful fight. Three fighting dogs opposed him. Turk, a purebred Bull and Terrier of a brown color, weighed about 16 kilograms. Shortly before, it had defeated and killed a much larger dog. From this fight the back of the head was still strongly swollen and nearly scalped. The second adversary was a fawn-colored bitch and the third was Tiger, a large black Bull and Terrier. The dogs were simultaneously set on the lion and all flew toward the lion's head without the slightest hesitation. After about five minutes, the fawn-colored bitch had to be removed, lame and apparently completely exhausted. After a further two minutes the second dog, Tiger, crawled out of the cage, dreadfully mauled by the lion. The brown dog, Turk, the lightest of the three, but of unbelievable bravery, carried on the fight alone. The spectators were treated to an extraordinary drama. The dog, entirely dependent upon itself, opposed by a beast that outweighed it approximately twenty-fold, continued to fight with totally undiminished courage.

Bull Baiting: Henry Alken, London circa 1820.

Although it bled from numerous wounds caused by the lion's claws, it seized the beast's nose and bit it at least six times. When the lion finally freed itself with a desperate effort, it put its full weight on the dog and held it between the front paws for more than a minute. During this time the lion could have mauled the terrier's head a hundred times, but did nothing at all to injure the dog. Poor Turk was then removed by its

III. THE PURPOSES OF FIGHTING DOGS

keeper, badly injured, but still alive. And in the moment that it was pulled from under the lion, it bit the lion's nose probably for the twentieth time.

The second day of combat with the lion Wallace on July 30, 1825 in Warwick ended in a clear victory for the lion, which was scarcely injured by the dogs. These fights are illuminated by an illustration from Pierce Egan's.Anecdotes from the year 1827. Here the fight between Wallace and his brave adversaries is impressively reproduced.

And with this account I conclude this chapter on the fights between our brave dogs and large predators.

4. FIGHTS AGAINST BULLS

When the Romans landed in Britain, they encountered the broad-mouthed, powerful dog Britannia. The Romans admired them greatly, because in fierceness and aggressiveness they exceeded by far the Molossus dogs known to them. The Roman garrison ordered several officers to obtain as many of the British mastiffs as possible and to bring them to Rome. Here the dogs were sent to fight in the large combat arenas against lions, bears, bulls and elephants, as well as

A few real fanciers: Henry Alken, London 1820.

III. THE PURPOSES OF FIGHTING DOGS

Wild boar hunt. Nymphenburg porcelain, circa 1800.

two-legged gladiators. Roman historians praised these British dogs particularly for their ability to break the steers' necks in the fights in the Roman circus.

How such exhibition fights against powerful fighting steers proceeded is shown by a superb illustration by Stradanus (1578). Here heavily armed men, accompanied by their large fighting dogs, and riders with lances set out against the bulls. From this we get the impression that the tormented bulls sold their lives dearly.

In Britain we find the first historical traces of bull baiting in the time of the regency of King John. After the wild aurochs was extirpated from the English forests, the nobility found a new adversary for their dogs in specially bred and enraged steers.

Because the bull is particularly aggressive by nature, it appeared a welcome pastime to the nobility to test the keenness of their dogs on these animals.

The first historical documentation is in the account *Survey of Stamford*: "William, Earl Warren, Lord of the town in the reign of King John (1199-1216 A.D.), *standing upon the walls of his castle at Stamford, saw two bulls fighting for a cow in the castle, till all the butchers' dogs pursued one of the bulls, which was maddened by the noise and multitude, through the town. This so pleased the Earl that he gave the castle meadow where the bulls combat began, for a common to the butchers of the town after the first grass was mowed, on condition that they should find a mad bull on a day six weeks before Christmas*

III. THE PURPOSES OF FIGHTING DOGS

for the continuance of that sport forever.

In the reign of King John, such fights were held in large arenas, where the bull ran free, unimpeded by chain or rope. For these fights, large dogs of sufficient size were brought. They were capable of pinning the attacking bull with the wildness of the tiger by the nose and to pull it down through their own strength, agility, and fierce determination alone. In these fights it was possible for the dog to knock the bull down through the force of its charge alone. We find this type of fight with free bulls, however, in only few accounts. Apparently there was a shortage of secure arenas, in which these fights could be staged without endangering the spectators.

Far more useful was the fight against the chained bull that was fastened by a rope to the bull ring. In this way fights could be staged in practically any town on the *bullanger*. The dog's goal in the attack was to pin and mercilessly hold onto the bull's nose. The bull's nose is its most sensitive spot; if the dog grips tightly here, the bull is virtually helpless. The bull, which was fastened to the bull ring, lowered its head as much as possible in the direction of the attacking dog. Experienced fighting bulls (game bulls) scraped hollows with their front hooves in which they hid the sensitive and threatened nostrils from the dog's attack. Only the horns greeted the attacker. The dog, in turn, had to keep its body as low as possible when attacking the tied bull, to expose as little area as possible to the bull's horns (*to play low*). Larger dogs crawled toward the bull on their belly. In the course of time, the dog breed was transformed. In the form of the old English Bulldog it was bred specifically for this fight: low-slung, but strong and brave enough to handle the bull. This new breed was extremely compact, broad, and muscular. It was about 42 centimeters high and weighed about 20 to 25 kilograms. Particularly characteristic of this breed was the

Wild boar hunt. Hunting book of Wolfgang Birkner the Younger, plate 30, circa 1600.

III. THE PURPOSES OF FIGHTING DOGS

Bull Baiting. If you go near Master George, he will pink you. Henry Alken, London 1823.

lower jaw, which projected considerably in front of the upper jaw. This made possible the strong, viselike grip. The nose was deeply set, which allowed the dog to get enough air as it gripped the bull. The old engravings of the time show the quite uniform type of this breed, although we must keep in mind that Henry Alken usually depicted the Bull and Terrier breed, the cross between the bulldog and terrier. These dogs were bred to attack only the bull's head. A Bulldog that gripped the bull in any other place was considered to be impure and not belonging to the breed. By attacking the head exclusively, the dog's teeth would not tear the bull's valuable meat and hide.

From the reign of King William III (1689-1703) we have a quite graphic account of a fight from the pen of the French barrister, Misson, who gives the following account of the bull fights: *They tie a rope around the base of the bull's horns and tie the other end of the rope to an iron ring, which is then fastened to a stake rammed into the ground. This rope, about five meters long, limits the bull's movements to a circle about ten meters in diameter.*

Several butchers or other gentlemen who wished to test their dogs stand around in a circle, each holding his dog firmly by the ears. When the sport begins, they let a dog go. The dog starts at the bull, the bull stands motionless and stares angrily at the dog and only turns its horns slightly toward it to prevent it from getting close. This does not influence the dog in the least. It again circles and tries to get under the bull's body.

The bull maintains its defensive posture, stamps the ground with its hooves, and places them as close together as possible. The bull's main goal is not to gore the dog with the tip of the horn (if the horns are too sharp, they are put in a type of wooden sheath); rather, the bull tries to push its horns under the body of the dog, which is creeping toward it as close to the ground as possible, and then to toss the dog so high in the air that it could break its neck in the fall.

To decrease the danger of a fall, the dog's friends are ready to break the fall of the dog's body with their own backs. Others stand ready with long sticks, which are held crosswise, with the intention of breaking the fall by letting the dog's body slide on the long sticks to the ground.

Despite all protective measures, such a ride in the air causes the dog to sing a quite vile melody and its face to break into a pitiful grimace. If it is not completely stunned in the fall, it will certainly creep toward the bull again, come what may.

Sometimes a second ride in the air finishes the dog off. Sometimes, however, it sinks its teeth into its foe. Once it takes hold with its fangs, it holds on as tightly as a leech, and would rather die than loosen its grip. Then the bull bellows loudly, rears up and lashes out, does anything to shake off the dog. In the end the dog either tears the piece of flesh that it has hold of from the bull and drops down, or it remains hanging with an

Bull Baiting. Now Captain Lad, Stafford, 1799.

III. THE PURPOSES OF FIGHTING DOGS

Aurochs hunt. Meissen porcelain, J. J. Kandler, circa 1760.

unrelenting, unending stubbornness, until the men pull it off. It would make no sense at all to call the dog off. You could strike it a hundred times, it would take no notice. Yes, you could even cut it to pieces, one limb after another, and it would still not let go. So, what is to be done? Some men must hold the bull tight and others drive a crowbar in the dog's mouth, to break the grip with raw force. So says our French jurist.

I must note, however, that the opening of the fight by the dog he describes sounds quite unusual compared to other accounts. Most other accounts report that the dog does not first circle the bull, but creeps directly toward it to grip the nose from below.

There were two different *rules of the game*. In a normal fight, the rule *let go* applied. In this case only one dog was the bull's adversary. Only after its victory or defeat was a second dog allowed *to try its luck*. In this fight two or three dogs were set on the bull at the same time, which of course drastically lowered the bull's chances.

We can understand the events of this time only by understanding the social conditions. We must see the sociological backdrop. What was considered entertainment at the royal courts, as a special honor for foreign guests, was carried over to the common folk through the commercial animal fights. The authorities have always understood that they must give the commoners *panem et circenses*, bread and games, to take their minds off their miserable social position.

With the bull fights, the nobility granted the commoners a very special entertainment. George Staverton, for example, stipulated in his will of May 15, 1661, that the income from one of his rental properties in Staines (Middlesex) should always be used to buy a bull each year. This bull was then set on by dogs, killed, and finally its meat was distributed among the poor. This benefactor of the poor also ordered that the money collected at this spectacle be used to buy shoes and stockings for the children of the poor. It is fairly certain that this honorable donor never considered the cruelties that would be suffered each year for the sake of his good deeds to the poor.

Related to this there is a peculiar story. In England people firmly believed for centuries that the meat of a bull set on by dogs was much more tender and tasty than from a bull that was simply slaughtered. Apparently, people had the idea that because of the excitement and exertions, the bull's blood *boiled*, which made the meat all the more tasty. A typical English saying bears witness to this superstition: *Trust me, I have a conscience as tender as a steak from a baited bull!* From our modern point of view it is unbelievable but true that in various parts of England there were laws that prohibited a butcher from slaughtering a bull that had not been set on by dogs beforehand. A whole series of judgments were handed down, in which the butchers were convicted of breaking this law, and were fined a substantial amount of money.

Additionally, the everyday existence of the lower classes in the Middle Ages was wretched and gray. The commoners had little education and training. The animal fights provided a welcome diversion. Moreover, bets could be placed on the outcome of the fights. To this was added the opportunities available to the owner of a dog. The butchers, who previously had used their dogs only to herd cattle, to break the resistance of unwilling animals for slaughter, all at once they could use their dogs — if

The pure-bred bull: lithograph, Gotha 1829, Carl Hellfarth's lithographic studio.

III. THE PURPOSES OF FIGHTING DOGS

Ox and bear fighting in Nurnberg: from the Nüremberg city library.

they were good enough — in the public fight against the bulls. Success led to fame and respect.

A sketch by Henry Alken from the year 1817 shows us very graphically how festively clothed people rode to the bull fights in carriages in four with the whole family or in the elegant one-horse carriages. The coach became the *platform of honor* at the presentation. No less impressive is the portrayal by an unknown amateur, *Bull Broke Loose,* from about the year 1800.

Anyone who examines these depictions of bulls, dogs, and people will understand that these fights were a public festival, a crystallization point of the social life in a time in which neither television nor soccer had yet stamped the human life. A fitting complement to this is another study by Alkens, *A Few Real Fanciers,* from the year 1820, doubtless a kind of parody of the fanatics of this *sport.*

That the church also promoted this bloody, cruel sport more than it condemned it is shown by the tradition of "bull running" in Tutbury, Staffordshire. Here a festival, the bull running, was introduced jointly by the Duke of Lancaster, John O'Gaunt (1340-1399), and the church. It appears that first there was a minstrel competition (the Minstrel's Court) and that the bull running was the crowning finale of the festival. It was the duty of the spiritual leader of Tutbury to award a bull to the competing minstrels at the conclusion of the minstrels' competition. Then the following took place: *After the feast, all the minstrels came to the gate of the priory in Tutbury to await the bull, which the steward of the estate of the priory was to deliver. The minstrels were unarmed. The tips of the bull's horns were sawed off and the ears and tail were cut off beforehand. Its whole body was rubbed with soft soap and pepper was blown into its nostrils. After that the master of ceremonies proclaimed that all but the minstrels must stay out of the bull's way. No one was permitted to get closer than forty feet from the bull or to obstruct the minstrels in their pursuit of the bull. Following this proclamation, the steward led the bull out the gate. If one of the minstrels was able*

III. THE PURPOSES OF FIGHTING DOGS

Wild boar hunt, Abraham Hondius, 1635.

III. THE PURPOSES OF FIGHTING DOGS

to cut a piece of hide from the bull before the bull reached Derbyshire, this minstrel was recognized as the 'King of Music,' and the bull was his. If, however, the bull reached Derbyshire healthy and without having lost a piece of hide, then it returned to the priory's possession. If the bull was captured and a piece of its hide was cut out, however, it was brought to the steward's house, a collar and rope was put on it, and it was led in a triumphant march to the bull ring in the High Street of Tutbury. Here it was attacked by dogs. The first fight was in honor of the King of Music, the second in honor of the Prior, the third in honor of the town. If further fights took place, they were in honor of the spectators. At the end of the fight against the dogs the bull belonged to the King of Music, and he could do with it as he saw fit.

I gladly leave it to the reader's imagination to picture this *festival*. The soap-smeared bull, with sawed-off horns and cut-off ears and tail, nearly mad with rage from this treatment and the sharp pepper, the pursuers chasing it through the town and fields, sliding off it because of the smeared-on soap, trampled by the raging animal, yet the pursuers still trying, despite the soap, to get a grip on it. Bull running in Tutbury was first mentioned historically in 1374 and was discontinued in 1778, thus after more than 300 years. Three hundred years of cruelty to animals — and all that to crown the King of Music! A more grotesque juxtaposition of two areas of human life is scarcely imaginable. We should not forget that these lower-class entertainments and passions that ran free here also awakened considerable commercial interest, as we have already seen with bear baiting. The simple steer that was presented by the butchers came to be replaced by an experienced and tested game bull. These bulls traveled with their owners

Bull-Baiting: S. Howitt, circa 1800.

III. THE PURPOSES OF FIGHTING DOGS

The Bull and Mastiff: Samuel Howitt, engraving, London, September 1810.

throughout the land. For a fee of fifty pence or a shilling, any dog owner could now test the courage of his dog against these professionally kept bulls. And what dog owner who had a true Bulldog or Bull and Terrier, what dog owner could resist the opportunity to show to himself and particularly his fellow citizens what his dog was made of? There is an account of a black bull from Tettenhall, which made a long round trip through the Black Country, the center of English coal mining. After this tour, the canal around Wednesfield and Wittenhall supposedly was filled with dogs' bodies for miles, *the trophies of the skill of the black bull of Tettenhall.*

On the same subject I present a quite colorful description of the fights from a letter to *Sporting Magazine* 1821, which was published under the title Bull-Baiting at Bristol. *Dear Sir, A large number of the fifty thousand readers of* Sporting Magazine *must have strolled over the incomparable lawns of Durdham Downs, . . . This was surely the prettiest place in the whole world for a bull-bait. A few minutes after our arrival the bull was led into the ring and tied to the stake. It carefully walked around its circle at the limit of the rope. It then circled closer and closer, and finally it took up its position in the middle of the circle. Occasionally it swatted its flanks with its tail. It sullenly stamped with its front hooves, as if impatiently awaiting the start of the fight.*

A nice two-year-old dog was the first to be set on it. It flew like a flash at the bull's head and tried with a daring attack to catch hold. The bull, however, with its horned head held close to the ground, made a short, but extraordinarily quick turn, thereby loosening the dog's grip in the moment the jaws snapped shut, and tossed the dog

III. THE PURPOSES OF FIGHTING DOGS

high in the air. It landed only a few steps from the bull's fetlocks and immediately crept undetected under the bull's body and seized it by the downward hanging lower lip. The bull had looked to see where its adversary had fallen and was completely surprised when it felt the dog's teeth bury themselves anew in its flesh, and on top of that in a place where it had least expected it. In the blink of an eye, however, the dog was again thrown from the ring by another desperate toss of the bull's head. The brave fellow, however, incited by the many shouts of the spectators, again crept toward the bull's deadly horns. The bull met it half way, and through another lightning-fast turn of the head stuck a horn deep into the dog's shoulder. It would certainly have killed this brave fighter in its rage, had the dog not been rescued by distracting the bull with another dog. The remainder of the drama now took a splendid course. The head of the old bull seemed barely to move, while he tossed two to three dozen dogs in a row into the air, to the obvious dismay of the dog owners, who often tried to catch the somersaulting dogs in their arms. Several of these brave dogs were injured repeatedly by the sharp horns and the wild ride in the air. Nonetheless, they crept anew toward their tormentor and came upon its hooves, where they were nearly trampled to death in their attempt to grip the head of their fearsome adversary.

Frederick W. Hackwood provides a very impressive account that shows to what extent the public life of a small town was influenced by only one of these fights. It should be kept in mind that certain Spanish influences were already detectable here on the English

The Grand Duke Nicholas, witnessing a let-go match by the dogs, for a Silver Collar at Oliver's Game Bull, at Combe Warren. London, December 1825, by Knight & Lacey. Designed by Theodore Lane.

III. THE PURPOSES OF FIGHTING DOGS

Aurochs hunt. Oil painting from the eighteenth century, unsigned.

scene. This, however, only serves to make the picture that much more colorful. The site was Bilston in the year 1743. A festival was organized here which drew spectators from the whole countryside. *Each member of the community had a set duty to perform. The town crier, dressed in a new uniform with a nicely powdered wig and a coat embroidered in silver, led the procession that festively accompanied the bull. The bull itself was handsomely adorned with ribbons and garlands. The noble steer itself was led personally by the awful Jack Willet. Jack was dressed in his old military uniform and was outfitted with an immense Spanish sword. He was marshal and master of ceremonies, the big organizer of the entire proceedings. The band consisted of a fiddler with a wooden leg and an asthmatic flutist, supported by the clacking sticks of the morris dancers of the town. The noise created in this way drew curious onlookers from miles away throughout the countryside.*

As the local historian reports, *an overwhelming sight met the eye. In the center of the circle stood the four-legged gladiator. The bull stared sullenly at all the unfriendly faces, particularly at the hostile-looking owners of the scarcely less vicious-looking Bulldogs. The dogs' owners were arrayed inside the ring, ready to set loose the dogs when the signal came to attack. At the entrance to the ring stood Jack Willet with drawn sword, which he waved back and forth to keep the spectators from crowding inside the ring. The mass of spectators gathered outside the ring. Because of the favorable lay of the land, they could stand in several rows, with the more distant rows standing above the closer ones, thus creating the impression of a Spanish arena. After all the preparations were complete, the town crier strode into the center of the ring and made the following announcement in a loud voice: All who want to try a dog, must pay five shillings. If the dogs are withdrawn, then for the same price anyone can enter the ring with a club. Whoever is able to bring the bull to its knees with the club, he wins rights to one-fourth of the animal!*

Now the sign was given for two dogs to attack. The bull met them with its horns, and with the first thrust it tossed one of them high in the air with entrails hanging out. A second dog took up position, and the two attacked the bull, until the same fate befell one of them. The people caught the dog as it fell, as if it were a child. Now Jack Willet stepped between the combatants and injured the bull slightly with the tip of his sword, which so enraged the bull that it began to foam from the nostrils. One dog after another was dispatched, until finally Shot, a dog whose fame was proclaimed throughout the land by the minstrels, was set on the bull. In an instant it had gripped the bull firmly by the nose. Jack Willet knew very well that nothing in the world could force the dog to loosen its grip, and he feared that this would bring the sport to an end much too quickly. So he cut off the flesh in cold blood with his sword. The dog fell to the ground and held the trophy of its skill in an iron grip between its jaws.

Now the town crier commanded that the dogs be removed from the fight, and they were replaced by a noble representative of mankind, armed with a club. In his advance on the bull, which had now gotten back its old strength because of the rest, the first man hesitated a moment, as if he had no idea of how to proceed. Finally, he raised the

Bull Baiting I: Henry Alken, London 1820.

III. THE PURPOSES OF FIGHTING DOGS

Bull Baiting II: Henry Alken, London 1820.

club carefully and suddenly swung it at the bull, in hopes of stunning it. The animal, however, well aware of the danger, quickly pivoted to the side and just as quickly stepped back and avoided the intended blow. Before the man recovered from the force of his own effort, the bull caught him with a horn and crushed him against the wall. Now another man picked up the weapon and smashed it against the animal, splintering one of its horns. The bull bellowed and stumbled in pain, but stood up again, rushed forward, freed itself from the chain, and set off through the middle of the crowd.

The air was full of the cries from a thousand throats. Make way, make way! they shouted. They ran away as fast as possible and left it to Jack Willet to live or to die there, as he wished. The scene was dominated by extreme fear. Men trampled women, women trampled children, and children trampled one another, all in the wild attempt to escape.

The hue and cry rang loudly and piercingly, with heavy oaths mixed with the desperate cries. Jack Willet alone remained calm during all this excitement. He leapt over the barrier and waited for his opportunity. The bull, weaker because of the great loss of blood, which streamed over its head and body, made one more feeble attempt to escape, fell to the ground, and was finally killed by Willet.

Is it necessary to describe in detail the scenes at the close of the day? The sale of the meat brought the necessary money for the closing feast. It was an orgy of gorging and guzzling, which Bilston never experienced before or since.

I have consciously included a full description of this scene in this book, because it shows how infinitely cruel man can be, with no consideration of the pain and suffering of the animal or the safety of his fellow man. The animals themselves are not cruel: they behave according to their instincts or the training that we humans have

III. THE PURPOSES OF FIGHTING DOGS

The Dog Pit. Dog Fight, 1979, United States.

III. THE PURPOSES OF FIGHTING DOGS

given them. Man is solely responsible for these degeneracies. The animals, bulls and dogs, deserve our compassion and our admiration for their courage.

Bull baiting degenerated more and more. There are countless accounts of the cruelty against bulls and dogs. The front hooves of the bulls were chopped off, to see how they could ward off the dogs on their bloody stumps. If they were exhausted from the fight, boiling oil was poured in their ears to make them lively again. Salt was rubbed into their wounds, pepper was put in their nostrils, and fires were lit under them to force them back onto their feet. Sadists can scarcely dream of worse than the chronicles report. Of the deluded dog owners themselves, we hear that they maimed their own dogs, which were stubbornly gripping the bulls, to prove that they were "real" Bulldogs, and to better sell their offspring. The senselessness of the dog owners who set their inexperienced dogs against far superior, experienced fighting bulls out of greed or to show off, and the greed and unscrupulousness of the bull owners produce a picture that must not be disguised by the prettifying and totally misleading word *sport*. This is a story of human degeneracy, the unparalleled cruelty against animals.

Bull baiting raged for a long time among bulls and dogs. The outbreak of cholera in Britain in 1832 led to a change in public opinion in Britain. In this pestilence, which wiped out entire villages, the people saw a kind of divine judgment. Even so, the bloody games continued. The bulls, adorned with garlands and ribbons, accompanied by howling and drunken followers of this sport, were led to the fights through the half-empty streets and towns. Even on Sundays, these sorts of Bacchanalian processions passed through the mourning towns. Thus, this movement increasingly was at odds with the feelings of the majority of citizens. The circles that had always sought to put an end to these "games" in the animals' interests increasingly won over the public opinion. Finally they succeeded in getting Parliament to ban the fights.

Despite the totally justified moral condemnation of the degenerate practice of bull baiting, the fact remains that our fighting dog breeds were substantially shaped by these fights. Indeed, they were specifically bred for this purpose! We can understand our dogs only from their history, from the purpose for which they were bred. The fighting dogs had a very hard time even surviving after the fights were banned. The Bulldog, for example, had lost completely its original purpose. In its further breeding this in turn led to degeneracy and faulty development.

Now, we should not put all the blame for this degeneracy on Britain. Accounts exist of fights between dogs and bulls from the Middle Ages up to the nineteenth century from many other countries, although not to the extent as from Britain. Nonetheless, there is very little doubt that atrocious degeneracies and cruelty to animals also took place in all these other countries.

Let us first turn to another portrayal by Stradanus (1578). Here we find a medieval depiction of five cattle, bulls and cows, opposing riders armed with lances, spearmen, and large dogs. The text states that the oxen were full of bellicosity, ferocity, and cruelty, and were exhausted by the barking and

III. THE PURPOSES OF FIGHTING DOGS 79

Hunting wild cattle, Stradanus (1523-1605).

attacks of the dogs, and after that were killed by the men. Flemming reports in his book *The Complete German Hunter* (1719) of the Brabant Bull Biter: *Where bears are rare, some gentlemen hunt steers, oxen, or bulls, which is a practice more suited to the butcher than the hunter . . . Yet in Brabant I have seen that the steer is tied to a long rope so that it can move in a circle and is set on by such dogs that bite the nose or throat. Because they, as previously reported, have a very strong bite, they are in the practice of hanging on without moving for a long time, . . . These dogs usually have short noses and are black around the mouth, the lower lip protrudes, their color is striped with yellowish or brown, and they have very unfriendly and bad-tempered eyes.*

We are indebted to the new art gallery for the youth, Gotha 1829, for a lithograph *The Purebred Ox*. It shows the intimate portrayal of the bull in defensive position against the typical Bulldog of its time, rather clear proof that this custom was also practiced on the European continent. These few examples of bull baiting on the Continent should suffice.

Many excellent depictions of the theme of bull baiting exist, so it was possible to illustrate this section with a number of them. First there is a very nice porcelain group from Staffordshire. It shows *Bull-Baiting 1791 Captain Lad*. Particularly impressive are the proportions of the powerful bull and the compact Bulldog. Captain Lad appears to have played the role of the organizer of the fight here. Henry Alken, to whom we are indebted for a whole series of illustrations of the cruel sport, of

Hate at first sight. Dog Fight, 1979, United States.

III. THE PURPOSES OF FIGHTING DOGS

course paid special attention to bull baiting. Besides the previously mentioned portraits, I present two matching illustrations called *Bull-Baiting* from the year 1820. The first shows the bull and dogs in attack position and the second the bull warding off the attack. We see how the owner of the white dog tries to break the dog's fall with his own back. Another print by Alken is quite impressive. This picture, *Bull-Baiting*, is not dated, but shows particularly well the power of the enraged bull and the danger it poses to dogs, owners, and spectators. Sir Edwin Landseer, the great English animal painter, apparently could not resist the lure of these fights either. I present two of his works: *Baiting the Bull* (1810) and *Bull Attacked by Dogs* (1821). I conclude this chapter with these impressive pictures.

5. DOGS FIGHTING DOGS

The dog fight stands at the center of all fighting sports. This kind of sport found the widest geographic distribution, and according to reliable reports, it still exists today.

Anyone who closely observes his fellow men and dog fanciers, in particular, will find, even today, dog fanciers who appear downright obsessed with the idea that the value of a dog is in its unconditional

Brabant Bull Biter. Bronze by Bosch based on an old terra-cotta.

III. THE PURPOSES OF FIGHTING DOGS

keenness and toughness. I am repeatedly asked where a dog can be bought that is one-hundred-percent tough and ferocious. They want just this sort of dog, no other.

Pull up to the bar and pay attention to the accounts of the proud dog owners. Baron von Munchhausen often sits at the bar, too. Here too you will hear stories from which it will be easy to conclude that the dog's only value for the storyteller is in the trail of blood the dog leaves behind in its encounters and fights with two- or four-legged adversaries. Are you in the Middle Ages, the eighteenth or nineteenth centuries? No, in the last quarter of the twentieth century! Pete Sparks — American, ardent admirer of the "cruel sport," and publisher of several historical reprints on dog fights —writes the following in the year 1974: *Dog fights, whose heyday was in the Roaring Forties (of the twentieth century!), still exist today only in the underground. Human society tries in every way imaginable to prosecute those responsible for such events, which they call cruelty to animals. But I am certain that, as long as there are still two real men with the right dogs, there will also be dog fights!* So writes Sparks in 1974.

Now let us take a step back. Captain L. Fitz Barnard, a famous heavyweight boxer writes about fifty years earlier: *Dog fights are not carried out in all countries, namely on grounds that the dog is such a faithful, loving friend and we hate to see it hurt. These grounds are lies, yes even worse than lies. We do not let a brave animal fight simply because we want to protect our own feelings. The dog loves the fight, but — as usual — we think only of ourselves. Dog fights are not cruel; no properly arranged fight can be cruel! Yes, a fight is certainly far less cruel than to travel with a dog to a dog show, without asking the question at all of whether the dog would not much prefer to fight than to be stuck in a show box all day long . . . If you have never had your own real fighting dog, then you can never know what a dog really is! The poor cur, leashed and dragged around dog shows, is good for nothing else in the world than to be looked at and to be made money from. The tame slave that helps us to*

Dutch Bull Biter from Flemming.

III. THE PURPOSES OF FIGHTING DOGS

Bull attacked by Dogs: Sir Edwin Landseer, 1821.

III. THE PURPOSES OF FIGHTING DOGS

Baiting the Bull: Sir Edwin Landseer, 1810.

hunt may be worth our notice; the dangerous pest, which runs behind the horse's hooves and barks, is of no use in life; but the fighting dog, with its rarely given deep love and its fearlessness in the face of death, has a right to love and respect!

Barnard, an expert in all kinds of fighting sports including cock fighting, characterizes the fighting dog as follows: *The fighting dog is the bravest creature in the world — not even the fighting cock is as brave. It fights under the most difficult conditions imaginable. When its turn comes, it must cross the pit and fight, or it has lost. We expect this from no other animal. It is prepared to fight anything, and when I say 'anything,' I mean 'anything': a piece of wood or a man, a fly or an elephant. Nothing stops it except for death. Its courage is not normal, and I do not believe that we can explain it mechanically. Courage comes from the brain, from the character, not from the body.*

I will discuss the question of cruelty against our fighting dogs at the end of this chapter, when we have learned more details about it. For the moment we can safely leave the views of Sparks and Fitz Barnard unanswered.

From contemporary authors, we learn that dog fighting was popular in England at about the beginning of the eighteenth century, but was still overshadowed by far by bull baiting in the eighteenth century. At the start of

III. THE PURPOSES OF FIGHTING DOGS

The treadmill. Dog training 1979 in the United States.

Bull Baiting by F. Barlow, late seventeenth century.

the nineteenth century, the dog fight was a permanent fixture at all events. At the turn of the nineteenth century, the transformation of the dogs began when terrier blood was crossed into the Bulldog breed. What was wanted was a faster dog that was more agile in fighting. The unconditional gripping, as in bull baiting, no longer counted; the dog fight could not be decided in this way. From the fighting Bulldog there arose a newer gladiator, the Bull and Terrier, and this also marks the beginning of the heyday of dog fighting in England.

The official ban of all fights in the year 1835 by the English Parliament actually served to promote dog fighting in England. Because of the small amount of space required for the pit, it was very difficult to enforce the ban. Bear baiting and bull baiting disappeared relatively quickly. Their followers were drawn to dog fighting, so that the heyday of dog fighting in England was between 1816 and 1860. After that time the police slowly began to eliminate the fights, but even at the turn of the twentieth century fighting dogs were still advertised for sale in English newspapers with a listing of all of their wins in dog fights.

In the meantime, fighting fever had spread to the New World. It began in about 1817, and dog fighting soon was just as widespread in the United States as in England. The *million dollar breed*, the *American Pit Bull Terrier*, did big business. To be sure, the Society for the Prevention of Cruelty to Animals in the United States forced an official ban in 1878, but in 1888 a big dog fight was still held weekly somewhere in the country. Later I will report about the big American championship fight in 1881 in Louisville. The centers of dog fighting in the United States were the cities of Baltimore, Chicago, St. Louis, Philadelphia and Boston.

The American Bull Terrier Club, which was founded in 1921, provides for *special competition for gameness* in its bylaws and also holds such competitions. Sparks says that the heyday of dog fights in the United States was in the 1940s. In 1976 the American Congress held public hearings on dog fighting, without coming to a concrete conclusion. Today this *sport* is still widely distributed in Latin America and Asia, as well as the United States.

So, we must conclude that in the 1990s numerous dog fights still take place in many countries, without animal protections societies being able to put a stop to them so far.

Now let us return to England at the turn of the nineteenth century. From our accounts of bull baiting and bear baiting we know that animal fights found a broad audience in the English population. Joy in the bravery of the animals, gambling and probably also the lust for cruelty — all connected with the social position of the population — these were the roots of the great popularity of these fights.

With the development of the new fighting dog, the Bull and Terrier, dog fighting soon gained in attraction. These dogs brought from the Bulldog the unconditional readiness to fight, insensitivity to pain, and from the terrier speed and a new fighting technique. This made the exhibition fights much more varied and thus more interesting to the spectator. For the dog breeders there were new chances to breed the great champion, to enter their dogs in new fights, and thereby to earn enormous sums.

From *Sporting Magazine* from the

III. THE PURPOSES OF FIGHTING DOGS

year 1825 I take the following account of a fight: *Dog Fight! On the evening of Tuesday, January 18, 1825, the 'Westminster-Pit' was filled to overflowing with dog fanciers from the capital. They wanted to see a fight between Bonny and the black newcomer, Gas, who was presented by his owner, Charley. The stake amounted to 40 sovereigns (about 100 dollars in our modern currency).*

All technical arrangements met the full satisfaction of the spectators. The ring was well lit by an elegant candelabra and a lavish array of candles. The dogs were presented for the fight in excellent condition at 8:00 p.m., led by their owners. Bonny was the 3:1 favorite and held these odds up to ten minutes before the start of the fight. This was proof of confidence in him, which rested solely on his generally known prior successes. For the nonpartisan spectators, the newcomer, Gas, showed much fire and aggressiveness. The duel lasted a total of one hour and fifty minutes. Then Bonny was carried unconscious from the ring, put in a warm bed, and bandaged immediately. Nearly three hundred persons had come to this fight.

This account shows the basic elements of the fight.

1. A **cash** stake from both dog owners. The victor wins the opponent's stake.

2. Betting. According to the reports of prior fights of the dogs and corresponding to the personal judgment of the individual gamblers, substantial wagers were placed on the dogs to win or lose. This increased considerably the nervous excitement of the spectators and was similar to modern betting on horses and Greyhounds.

3. Admission fee. Three hundred spectators came to the fight on January 18, 1825 at the Westminster-Pit. Admission fees were raised depending on the importance of the fights and mostly went to the owners of the pit, but could also in part go to the winner.

To gain a better understanding of the events, below are the rules of competition, such as were summarized in 1910 by Eugen Glass from old English fighting contracts.

Rules of competition for dog fights

1. Both dogs must be weighed before the fight. Neither of the dogs may exceed the agreed upon fighting weight, otherwise the stake is lost. In the pit, the time keepers, two umpires, and the referee are selected and named with the agreement of both parties. The pit should have dimensions of 12 x 12 feet and is divided diagonally by a white center stripe; the opposite corners of the ring are marked in the same way, whereby the ring corner should have a diameter of about 60 centimeters.

2. The dogs are to be licked with the tongue before the fight to determine if any dangerous substance or ointment has been applied to the dog. If this is the case, the dog must be washed until completely clean under the direction of the referee or the referee declares the violator's stake in the fight as forfeit. The dogs must be licked with the tongue before and after the fight if a challenge is made and the referee orders it. Two buckets of clean water must be provided for wiping off the dogs.

3. The corners are drawn for, and each of the dogs is released by the seconds in a fair manner. The dog that makes the first mistake, that is, the dog that turns away from its opponent first, subsequently is the first that must cross the center stripe to the opponent (to scratch), assuming that at the time of its first mistake or retreat its second picked it up. After the first separation, the dogs must cross the center line alternately, regardless of which one made the mistake or was the first to turn away. The dog that last crosses completely over the center line will be declared the winner. If a dog is so exhausted that it can no longer bite the other, but can still cross the center line, then the dog whose turn it is must cover the entire distance without stopping. To have crossed the center line fairly, all four feet must be on the other side of the center line.

4. None of the seconds may touch a dog or be unfair to a dog or to the other seconds across the pit. If he is unfair, his dog must be disqualified by the referee. It must be considered a *foul* if the second throws the dog across the pit or steps on the head of a dog or stamps his foot next to it. If the second has inadvertently picked up his dog, then he must put it down immediately, so that the fight can be carried out fairly. If the premature pickup was deliberate, it must be judged a *foul* by the referee.

5. To ensure a fair separation or the proper pickup of the dog, both dogs must be completely free of each other, both with the heads and with all four feet. Each dog's second has to watch very closely to make sure he picks up his dog only when, as described previously, it has moved away from the other dog. If while picking it up the other dog attacked again, the second must put down his dog and wait fairly the further course of the fight until the time of a proper pickup. This rule in particular must be

A Dog Fight. Rowlandson, print, London May 1, 1811.

III. THE PURPOSES OF FIGHTING DOGS

Henry Alken: Dog Fight in the Street, London, 1820.

followed to the letter. If there is a difference of opinion here, the umpires and the referee must be called in. The referee's decision, whether right or wrong, is binding in all cases. Only very experienced referees should take on such thankless duties and only with the agreement of both parties prior to the fight. After the dogs are picked up by the seconds, they go to their corner and wash and rinse the fangs of their dog. A minute is permitted for this; the timekeeper must shout after forty-five seconds, *Everyone out of the pit!* After fifty seconds he shouts, *Ready!* Each dog must be held in its corner so that the head is just across the stripe. After sixty seconds comes the cry, *Let go!* and the timekeeper turns to the particular dog owner, names the dog and the owner, and says *It's your round, start!*

6. Each party is allowed to place a silent observer in each corner with the agreement of both parties. They must stay at a respectful distance from the dog, water bucket, and so forth. If there is the suspicion of any sort of unfairness on either side, the decision lies with the referee.

Later rules of competition also regulate when the cash stakes are to be paid in by the dog owners, who was to hold the money, what would happen if the published time of the fight had to be delayed because of the intervention of the police, and so forth. We can see that these fights are organized down to the smallest detail because of the high stakes involved. Nothing is left to chance.

If we translate the above rules in plain English, we see that basically only dogs of the same weight compete against one another; it was strongly believed that with well trained dogs, differences in weight would give a decisive advantage to the heavier dog. In these fights, male dogs also fought against bitches. On this subject, Fitz Barnard writes: *Many people believe a male dog will not fight against a bitch. They don't know what they're talking about! A fighting dog fights a bitch just as readily as anything else. This fighting instinct is so strongly developed that you have to be very careful when you breed these dogs. They are undecided between love and fighting with a very decisive orientation toward fighting.*

Probably the most famous dog pit was the aforementioned Westminster-Pit. Spectators came from all parts of the country to visit this pit in the capital. Usually several fights were staged in an evening. Contemporary newspaper reporters tell of its unique atmosphere. Dogs, held tightly by their owners at the ring's edge, howled in rage, that they had not yet made their appearance, or in an even more uncanny silence stared at each other, their tongues licking their lips. A glance at Rowlandson's *A Dog Fight* gives us insight into the atmosphere of a big fight day. This print reveals not only the dogs' suffering, but especially the fanaticism of the crowd. A unique portrayal!

Such a fight could end in a few minutes, but many took hours. Approximately two hours is reported as the average time, but accounts also exist of fights lasting more than four or five hours. We can easily imagine that such physical exertions on the part of the dogs demanded not only a particularly tough and tenacious breed, but systematic training as well.

Famous fighting dogs were very valuable for their owners. One of the best known dogs was Belcher, the winner of hundreds of dog fights. He was owned by a famous prizefighter of the day. In a change in ownership among the professionals of those years, a Mr. Mellish paid twenty guineas for Belcher, a fortune at the beginning of the nineteenth century. After a series of other big wins, Mellish gave this dog to Lord Camelford and received for the dog the Lord's favorite gun and a pair of valuable pistols.

Two other famous dogs, Young Storm and Old Storm, both weighing 32 kilograms, were successful in all their fights. They were descended from the famous Paddington Strain and must be considered to be the first Bull and Terriers of predominantly uniform type. A portrait from *Sporting Magazine* of the year 1824 shows us the dogs. Of Old Storm it is reported that in two of his fights the adversary was left dead in the ring. Young Storm already won four dog fights, each lasting more than an hour, by his second birthday.

In 1819 a poster announced a fight in the Westminster-Pit with a stake of one hundred guineas. One of the dogs was *the famous white bitch from Paddington,* whose wonderful performances were so well known that nothing more need be said here. Her opponent was a brindled male from Cambridge, *a noteworthy and well-known favorite, which has already undergone quite intensive tests.* A stake of one hundred guineas! The fighting weights were twenty kilograms each.

Dog fights were so popular in these years that, if you did not know the name of a famous dog, you proved

III. THE PURPOSES OF FIGHTING DOGS

that your were not up to date on sporting matters. Jesse reports that a man, riding through Wednesbury, stopped at the toll gate. Then the church bells began to ring and he asked the reason for the chiming. *Old Sal has delivered!* was the answer. Still not making head or tail of it, the rider asked who Old Sal was. *Old Sal, Old Sal!* came the answer, as if speaking to a deaf man. *You really don't know who Old Sal is?* Then the toll collector explained that Old Sal was a fairly old, but very famous Bull bitch and that she had just given birth to her first pups. And the church bells carried these glad tidings across the land.

We can imagine that these dogs were extraordinarily valuable to their owners. The owners of fighting dogs loved their dogs. At least they were extraordinarily proud of them. The most promising pups of a litter were usually more diligently cared for and coddled than were their own children. In any case, it was always a very divided love. It was driven purely by expedience and strongly influenced by personal vanity and greed. The love for the animal disappeared when it was time to prepare it for its task. It is reported that the initially coddled pups, when they grew a little bigger, were put in darkened rooms, isolated from all other people and animals. Only contact with their own master, but not with their siblings or other dogs, was permitted. Also, contact with strangers was strictly prohibited.

Everything was directed toward the job in the pit. Before we turn to the essential training of the dogs for fighting, first let us focus on a big fight. For this fight an advertisement was placed seeking the best fighting dog in the United States. Louis Kreiger, the owner of the white imported male, Crib, from the city of Louisville, advertised in the *Police Gazette* a fight with his dog with a stake of 1000 dollars each by the two dog owners. This challenge was accepted by Charles Lloyd (Cockney Charley) from New York for his brindled male, Pilot, also an imported dog. The fighting weight for both dogs was 28 pounds. The fight took place on October 19, 1881 on Garr's Farm in Louisville. Two hundred spectators paid the entrance fee of a dollar to see the fight in the big barn in which the pit was set up. The old Cockney Charley marched through the crowd and offered "$25 on Pilot!" "$50 on Pilot!" "$100 on Pilot!" $3000 was bet, in addition to the $2000 put up by the two dog owners. When we consider the value of this money in the year 1881, we can understand the chances and risks involved in dog fights. With good reason, the Pit Bull Terrier was called the *million dollar breed.*

After the dogs were washed clean and tasted (licked), and then rubbed dry, they appeared in the ring. I have taken the following from the original account:

Hughes, Chief of the Louisville Fire Department, announced that at the request of the referee, the two dog handlers were to be searched. The dog handlers searched each other's clothing, to satisfy themselves that nothing was hidden that could injure their own dog. When this examination was over, at 9:20 a.m. the command was given to loose the dogs. Their covers and muzzles were quickly removed and the dogs were released.

Both growled deeply and with a wild charge Crib rushed to Pilot's corner and attacked his opponent. He seized Pilot by the nose, but the brindled dog shook him

III. THE PURPOSES OF FIGHTING DOGS

off and seized him by the right front paw. Then Pilot released his grip on Crib's foot to get a better grip on his throat. Crib was able to free himself and again seized Pilot by the nose, but suddenly let go and gripped Pilot on the back of the head and ear and threw him to the ground. Lying on the ground, Pilot seized Crib by the chest and applied a terrible bite. Because he could not get a proper hold here, however, he let go and gripped Crib by the left ear. Then Pilot again released his ear hold and put Crib's left front foot between his molars. He closed his jaws more and more tightly and the bones of Crib's leg crunched. This terrible pain seemed only to enrage the dog from Louisville even more, and with a tremendous effort with his ear hold he tossed Pilot to the ground five times in a row. Now Crib again seized Pilot by the nose, which consequently became his main point of attack, and again threw the dog from New York to the ground. Then he suddenly let go of Pilot's nose and began to gnaw on Pilot's front leg.

With the fighting style that Crib now showed, he appeared to the people from Louisville to be the sure winner of the fight. But now it was Pilot's turn to carry the fight, and the way in which he gnawed on Crib's left leg was terrible to see. Crib freed himself with a wild growl, before being seized immediately in the same way again. With another try Crib again gained his freedom and for a full five minutes the dogs fought with various ear holds until finally Pilot forced his way under Crib and — holding him by the ear — slammed his head repeatedly against the floor of the pit.

Crib now secured a hold on Pilot's

Dog Fight, Henry Alken, London 1824

III. THE PURPOSES OF FIGHTING DOGS

Young Storm and Old Storm: Bull and Terrier *Sporting Magazine*, October 30, 1824, London.

throat and — although he now fought on only three legs — he was able to throw his opponent to the ground. This again served only to incite Pilot. He threw down Crib with a throat hold and then seized a hind leg. Crib countered by taking hold of Pilot's leg. At this moment another $1000 bet was placed on Crib to win. This was accepted immediately by Cockney Charley, Pilot's owner.

The fight had now lasted 42 minutes. Crib now succeeded in escaping from under his opponent, but the poor fighter's courage had disappeared. He turned to the side of the pit and was ready to run out when he was seized again by the brindled dog and pulled back into the middle of the pit. At this moment Crib was a beaten dog, but Pilot was not yet satisfied with his victory. Rather, he was determined to kill his rival, now that he seemed to have the chance.

Crib again turned to the side of the pit, and this time he made it, with Pilot in hot pursuit. Now Pilot seized the lower jaw of the 'Louisville Lapdog' and refused to let go. By doing this he forced the handlers to pick up the dogs together and return them to the ring. Pilot tossed Crib into the corner with an ear hold and held him securely there. Kreiger vigorously fanned air to Crib with his hat, but this did not help the dog, which quickly lost its strength. From this moment on Pilot's only goal was to try to shake the little remaining life out of Crib.

The fight lasted exactly one hour and 25 minutes. Other than various severe bites on the head and shoulder, Pilot was only slightly injured.

How did you come up with the ideal fighting dog? It already began with the breeding of the pups. The winners of the big fights were bred. The blood had to be pure. Not only the two parents, but their ancestors, too, must have survived successfully the purgatory of the pit. We have here an example of unconditional selective breeding — based purely on performance over many dog generations. In this selection lies the secret of our fighting dog breeds for well over a century. The basic social order of the dog pack was systematically bred out of these dogs. The fighting dog accepted no gesture of submission, it had no inhibition against biting the bitch, the defeated rival. I must also mention that this change in the nature of the dog as a pack animal is accompanied by certain dangers. For example, we are rightly warned against holding an enraged fighting dog or carelessly getting in its way. When the animal is in full action on the other side of its stimulus threshold, it seeks out the fight. In this case the dog could — to be sure only in extreme cases — turn on its own master. The owner of a descendent of these old fighting dogs must always be aware of this danger.

We encounter analogous problems in breeding and in the behavior of the mother bitch toward the pups. It isn't without just reason that experienced breeders constantly watch over the mother and pups within the first weeks after birth. Here too, instincts from the fighting dog heritage of the mother could be triggered, which contradict the natural maternal instinct.

Even the pups of the fighting dog breeds show clear differences to the behavior of other pups. They fight in the nest completely differently from the pups of other breeds of dog. I myself have saved the life of a Bull Terrier pup, six weeks old, whose artery had been severed by one of its siblings. Once the little fellows have gotten a grip on one another, with one pup you can lift a cluster of five more pups from the nest. Each clings firmly to the fur of the other. We break up the cluster one by one by cutting off the air to each pup in succession. This heritage of fighting dogs is still unmistakable in the litters of fighting dog breeds today. Two pups, about three months old, can already fight each other for twenty minutes without interruption if we do not intervene. The fight, in proper fighting dog style, takes place without a sound, so that it is best for the breeder to separate the pups early to avoid unpleasant surprises. Only breeds with a strong infusion of terrier blood make sounds during the fight; the bulldog fights with *the silence of the grave.* It is said of an approximately four-month-old fighting dog that, at this age, it can already kill cats.

Dog fights of the described length, ferocity, and intensity required a carefully trained fighting dog. A wide variety of training methods was used. The exact training program was a closely guarded secret of the individual trainer. There are a number of accounts describing the different training programs used with fighting dogs. Below I will briefly present some of these cruel, individual training methods.

The yearling, for example, was chained to a wall by the chest harness or collar. Within the chain or rope a central section of rubber was inserted, which provided a certain degree of elasticity. Now, beyond the dog's reach, it was shown a rat in a trap or a cat. The dog repeatedly jumped at this prey, but could never quite reach it, because the rubber pulled it back. Experts recommended increasing the duration of this exercise from an initial five minutes

III. THE PURPOSES OF FIGHTING DOGS

up to an hour.

The *golden rule of a sportsman* was to incite the fighting dog with a mongrel dog. Such mongrels could be bought cheaply anywhere. It was essential that when the exercise was over that the dog be allowed to kill and eat the cur. To increase the effectiveness of this exercise, our experts recommended building a big fire after a few days so that the fighting dog would also have to fight against the heat of the fire. With this method it was possible to make a dog so fit within a few weeks that you could confidently bet $50 on it. It was also important, however, except during the daily training, to always keep the fighting dog in a completely closed, dark room and to feed it only raw meat and blood. Within about two weeks these dogs in the dark rooms were just right. Usually by the eleventh day they have already emerged like a tiger.

Other recipes for toughening up our fighting dogs can also be found in the relevant books. One highly recommended exercise is the jump rope. This rope is tied to a ring on the ceiling. A sack is now tied to the end of the rope, a cat put in the sack, with the cat's paws sticking out through four holes in the bottom of the sack. The cat tries to avoid the bites of the jumping dog and thereby starts the sack swinging. The dog feels the cat's claws and does everything in its power to jump up and reach the bundle. When the exercise is over it *gets the cat as a reward*. It is not hard to imagine that in this way you can soon induce the dog to hang by its teeth from any bundle hanging from the jump rope. The dog is then swung back and forth on the rope, thus strengthening the power of the jaws and neck and back musculature.

The dog trainers had available still another instrument of torture: the treadmill. This was similar to the treadmills used in old mills or wells, where the donkey or ox walked round and round to grind grain or pump water. The dog was trained on the treadmill several times a day, starting with five minutes per set and working up to an hour without rest. We found a caricature of such a treadmill exercise in Jack Meeks' *Memoirs of the Pit*, which not only shows the system but also caricatures the excesses that frequently occurred here. The principal goal of the work on the treadmill was not to reach the reported fighting weight, but to strengthen the heart and lungs. We took another illustration from the report in the August 1979 edition of *Geo* about training in the United States.

Of decisive importance for maintaining the correct fighting weight were long daily walks with the dog chained alongside the trainer. These walks were up to nine miles long. Whenever possible hard roads were used to toughen the paws. To this was added a systematic diet, again usually secret recipes of the individual trainers. The diet usually consisted of lean, muscle-building, concentrated food, mainly raw or cooked, lean beef mixed with bread crumbs and sometimes raw eggs.

It was generally believed that it took fours weeks to prepare a dog between the time the fight contract is signed and the fight. During this time the dog had to be brought down to its fighting weight and into proper condition through a combination of a specialized diet, intensive exercise, and systematic body building. Keep in mind that it had to be in good enough condition to survive a fight lasting even four or five hours. The previously described combination of the jump rope, treadmill, walks, and appropriate diet appears to be optimal for this purpose, although it would be best to forget about the senseless dark room.

It is also interesting that all experts recommend rubbing down the dog

III. THE PURPOSES OF FIGHTING DOGS

A Match at the Badger: *Why, Master George do you expect that little thing to draw the badger?* (detail) Henry Alken from the series Master George, London 1823.

vigorously following exercise and massaging it by hand. They recommend massage times of approximately a half hour every time the dog was exercised, and that the dog must be bedded on warm straw in a draft-free stall after any exercise. The intensive massaging was an important component of the body building.

It is believable when experts claim that the proper training of the fighting dog for the pit was more difficult than the preparation of a race horse for an important race. It was certainly pleasant for the dog that such intensive preparation always demanded close contact with the trainer. The dog and trainer became a team during the training period, which also enabled the trainer to inspire his dog to greater heights on fight day.

The million dollar breed naturally attracted many greedy, shady characters. Numerous precautions had to be taken to limit cheating and human cruelty in this *sport*. In the literature we repeatedly find advice on everything you must do to protect your own dog. Each trainer not only had to guard his own dog against hostile influences but had to put observers in his opponent's corner to make sure that nothing illegal took place there. The licking of the dog before the fight, either by an official taster of the fight committee for a correspondingly high fee, or by the dog owner himself, was done to prevent poison from being concealed in the dog's coat. You also had to keep a close watch on your opponent to make sure the he did not give your dog a secret injection of poison while washing the dog, which

III. THE PURPOSES OF FIGHTING DOGS

might not take effect for a half hour. Furthermore, something could be poured in the dog's mouth or ears. A slow-acting poison, such as strychnine, or bird shot in the ear were things that not even an experienced fighting dog could overcome.

Another very popular trick was to apply a mixture of cayenne pepper and aloe to vulnerable parts of your own dog after it was washed. This was the hottest mixture that you could possibly put in your mouth. No dog would bite these treated places. Bird shot in the ears, only a few pellets, makes the dog almost mad with pain. No doubt there were many other dirty tricks that were used to influence the outcome of the fights to the cheater's advantage.

Even during the fight there were maneuvers that sneaky seconds could play on the novice. If the rival dog was the first to cross the center line, you held your own dog between your legs with its head facing the opponent until just before the attack. If your own dog was white, you always wore white clothes, and you wore dark clothes with a dark dog. When the novice crossed the center line, its rival thus appeared to be giant sized. If it froze, then the second could pick up his dog and claim victory for his dog, because the other dog had not attacked immediately. The system of picking up the dogs during pauses in fighting usually was also taken advantage of by the experienced second, to the detriment of the novice.

How intentional or unintentional wrong decisions by referees could affect the outcome of fights is splendidly illustrated in the account of a true-to-style competition. E. S. Montgomery took the following description of a dog fight from a story in an American magazine. This fight, which took place in the year 1800 in New York City, should be of interest to my readers:

Paddy from New York and Trix from Philadelphia fought a fight to the death on the appointed day. The fight took place in East Town Hall. Paddy and Trix are of the Bull Terrier breed. The contracts for this fight had been signed weeks before. One of the conditions of this fight was that each party could not bring more than a half dozen spectators to the event. This condition was not met, however, because a medium-sized crowd had already found its way into the rather dilapidated building before the fight could begin. Most of the spectators were old men. Apparently a number of them had personally attended virtually all the big dog fights in the last half century. For them this was the greatest pleasure in the world, and their enthusiasm for past fights at times reached rather unbridled proportions. Especially during the most thrilling moments in the fight they gave the impression that they could barely restrain themselves from jumping into the combat arena and embracing the blood-soaked dogs. The wager was 300 dollars a side. The dogs weighed, as stated in the contract, 28 pounds each. A dog arena is never a particularly quiet place during a real dog fight, and this fight, of which we report, was no exception to this rule.

Paddy is a white-and-yellow Bull Terrier, about four years old, well known to anyone who enjoys dog fights. He can look back on many fights and victories.

Trix is — or, more precisely, was — 'one of the coming world champions.' That at least was the opinion of his owner in Philadelphia. He is just 24 months old and pure white. A truly good-looking animal, but his fighting qualities pale in comparison with his older opponent.

The dogs were washed and licked according to the rules. Dog fighters never trust each other, so both dogs were washed and cleaned in the same tub, to prevent one of the parties from pouring dangerous drugs or poison in the water. Paddy and Trix were cleaned very thoroughly and then each owner licked the rival dog with his own tongue. When it was determined that there was no poison in the dogs' coats, the ring was cleared. The mood of the crowd soared and the command *Get ready!* was given. Then came the command *go* and the eager dogs were released from their corners. They sprang at each other like a flash, each attempting to sink its teeth into its rival. The skill and cunning of Paddy was soon revealed. He showed that he was a crafty old dog. Truly, he shoved and bit his rival through the ring in such a terrible fashion that the whole crowd screamed loudly in joy. Trix was neither inactive, nor did he patiently endure the pain caused by Paddy's fangs boring into his flesh. On the contrary, he fought bravely. And he fought well, but, cheered on by his fans, Paddy steadily gained the upper hand. Soon Trix lay stretched out on the floor, his pretty coat soiled with dirt and covered with blood.

How his owner talked to him now, 'Trix, come on boy, tear his head off!' How Paddy's second yelled out, 'Paddy, tear the Quaker to pieces!'

The fight lasted two hours. Trix was finished in the best terrier fashion; that is, he was mauled terribly, and it was obvious that he had absolutely no chance of winning the fight. He was almost dead when the dogs were brought back to their corners, carefully washed with the sponge and then set upon each other again. It was Paddy's turn to attack, and this attack was one of the best that had ever been seen in the fighting ring. Then, to the utter amazement of all the spectators, the referee cried, 'This is an unfair attack!' Then he looked around as if seeking support and shouted, 'The attack was unfair! I hereby declare this fight a draw!'

Fifty spectators began to curse and abuse the referee with every name they knew from their many years of attending dog fights. 'You're a fine, thieving pig!' — 'You Philadelphia cheat!' — 'Bought scoundrel!,' to name only a few of the pet names that were directed toward the referee from the lips of the disappointed men. It would have taken only a spark for them to have fired the pistols that they waved around. The referee turned pale from the uproar that he had caused. He managed to find a few supporters, though. Finally he declared that all bets were off and left the ring in good health, which, all things considered, was quite amazing. Paddy had won his fight fairly, and he should have won the money. Trix was so badly wounded that he probably did not survive the fight.

'Now, if that wasn't the most deceitful maneuver I've ever seen, you can cut my head off!' — This was the comment of one of Paddy's disappointed supporters when the whole thing was over.

Now let us move forward to another time in the United States of America. From the German magazine *Stern* from the year 1974, we take the following account: *Blood sprays high up the ring wall, and you can hear bones splintering. Red Lady, still on the throat of Hungry Tiger, suddenly gives up. Her flanks heave wildly. Hungry Tiger has clamped his jaws on her right front leg. It is the 76th minute of the fight. The sports fans who stand ringside slowly get in the mood. Canned beer for the men, liqueur for the women, and the children are having fun without any alcohol at all.*

Red Lady, weighing 39 pounds, six years old, had actually been the favorite. The victorious veteran of 14 dog fights. She had bitten six opponents to death, five died from their wounds after the fight, three survived. Her owner had bet more than $1000 on her to win over Hungry Tiger. He now slides on his knees across the blood-smeared ring and encourages his Lady, who no longer can, with pleas and curses. 'Come on, my Baby, please, please!' and right after that, "Bite, you damned bastard, bite!'

The crowd howls as the dogs, wedged together, tumble through the twelve-foot ring. They tear each other to pieces. More cans of beer are handed out.

Then Red Lady gives up. With tail between her legs, her belly almost torn open,

III. THE PURPOSES OF FIGHTING DOGS

she limps on broken front legs to the corner of the ring and tries to escape Hungry Tiger by jumping over the knee-high boards. She cannot make the jump and remains hanging on her belly on top of the boards. The fight is interrupted, the dogs are taken to their corners.

If Red Lady's owner is unable to get the half-dead animal to fight in the next few minutes, the match will be over and the money will be lost.

They can barely hold back Hungry Tiger. On the referee's signal the dogs are set on each other again. Tiger attacks Lady viciously. Lady fights back with a final burst of energy. After three minutes Red Lady is finished. She bleeds to death in the ring.

Someone sticks a meat hook into the cadaver and throws it into a steel barrel positioned ten paces behind the row of spectators. Hungry Tiger — fighting weight 40 pounds, 4 years old — is the new star. Children surround him and watch him lick his terrible wounds. The adults say that they will bet on him in the future. Then the next competitors are announced.

With the kind permission of *Stern Magazine* and *GEO* we present a series of photographs from the August 1979 edition, which show, better than words ever could, how dog fights are played out before a fanatical, feverishly betting crowd. A repulsive, frightening sight! These pictures are necessary, however, to show this bloody reality to all who secretly sympathize with this kind of degeneracy! Dog owners and spectators — a real problem for the psychiatrist!

Dog fighting is banned in the United States only by the laws of the individual states. Successful dog fighters earn $25,000 and more a year. The biggest fine imposed so far was $875. A clear choice!

There exist credible accounts of dog fights from numerous civilized countries. Jack Macks, for example, collected more than a hundred detailed fighting accounts in the years between 1887 and 1939 in the United States, supplemented by numerous individual reports of famous fighting dogs. Thus, we can probably assume that this plague of fighting dogs has virtually been extinguished in the land of origin of the breeds, but that even now people still find pleasure in this bloody spectacle. I watched a television broadcast on the BBC 1 on October 23, 1980 in which the famous Staffordshire judge, Ken Bailey, confirmed to the reporters that even in 1980 secret dog fights were still being held throughout England.

In the discussion of bear baiting and bull baiting we saw that their supporters contended that these *sports* were merely tests of gameness. With all due respect to the social background in the nineteenth century, it is simply not true that the *sports fans* staged these fight to give the animals the chance to prove themselves. These dogs were developed through systematic breeding and robbed of their natural social behavior to carry out these fights for the sensation-seeking public. The sports fans were acting out of their own greed, not in the interest of the poor dogs.

These fights have given us the ever so interesting fighting dog breeds. It is our job to keep and train them so that their natural instincts do not degenerate but work to the benefit of man and dog.

6. FIGHTS WITH THE BADGER

There is scarcely an animal capable of defending itself on which the bravery of the dog was not put to the test in Old England. The badger always showed itself to be a dangerous opponent for any dog, so what would be more obvious than to haul it into the fighting arena as a true opponent of the fighting dog? The badger, weighing up to thirty-five pounds when fully grown, has an extraordinarily dangerous bite, which it is willing to use recklessly when threatened.

In order to use the badger's ability to defend itself to test the dog, artificial badger dens were built, captured badgers were put in them, and then the dog was set on the badger. In the *Encyclopedia of Rural Sports of the Year 1840*, Blaine describes this *sport* as follows: *In the Middle Ages badger baiting was practiced largely as it still is today . . . The badger was placed in an empty box, which was furnished in imitation of its den, and from there a tunnel led upward. The owner of the badger puts his animal in the box. Then a timekeeper is equipped with a watch and the badger's owner releases the dog for the fight. Whoever wants to pit his dog against the badger let it slide into the tunnel. Usually the dog is seized immediately by the badger, and the dog in turn grips the badger. Each bites the other and each pulls and tears at the other with all its might. The dog, however, is now quickly pulled out by the tail by its owner, with jaws clamped obstinately onto the badger. The two are separated and the badger is returned to its den. Then the dog is sent back in to seize the badger, and it is again drawn out with the badger. This scene is repeated again and again. The more often a dog is able to seize the badger within a minute so that both can be pulled out together, the more it is up to the task and is considered game.*

From other contemporary accounts we see a whole series of variations of this description of *drawing the badger*. In one account it is reported that before it is put in the box, the middle of the badger's tail is split so that a chain can be attached to it. Behind the box a stake is rammed into the ground and the chain fastened to it. This chain is so long that the badger can defend itself freely and can also be drawn out, but not long enough for it to escape. Often several dogs are set on the badger in succession. Here the individual time of each dog counts, not the number of times the badger is drawn. Here the dog's handler maintains contact with his dog throughout the fight by holding his dog tightly by the tail with his left hand, so he can pull out the dog and badger as quickly as possible after they clamp onto each other. To speed up the drawing out, the dog's owner bites his dog on the tail (a method that probably does little good with fighting dogs, since, as we know, they become almost insensitive to pain when enraged).

To illustrate this sport we look to two prints from Henry Alken. *A Match at the Badger* shows a dog owner, two timekeepers, and the badger's owner at the beginning of the fight. The second illustration gives us the picture after the badger has been drawn out and the dog is to be made to let go by a bite on the tail. Note the type of dog depicted by Alken in each print, clearly the Bull Terrier of the 1820s with a considerable infusion of terrier blood.

With badger baiting we can clearly see that something is demanded here from our fighting dogs that contradicts the other duties we have encountered so far. Up to now we have been familiar with the merciless fight to the bitter end. Here the dog must draw out the opponent several times within a minute, break off the fight, and immediately start anew. Normally, without the contact of the

III. THE PURPOSES OF FIGHTING DOGS

A Match at the Badger: Henry Alken, London 1820.

human hand, the fight would be carried out completely in the den. If the dog were left alone, the spectator would see nothing, as the fight would remain hidden underground.

We have mentioned previously that the badger has an extraordinarily strong bite. Clever and experienced dogs therefore went head first into the den instead of leading with their legs. In this way they could fight the badger face to face without putting their own legs at risk. Such fights underground are certainly known to hunters; on our hunting preserves they can last for hours and can also end in the death of the brave dog.

Hackwood reports on an unusual fight in Preston in the year 1817. Here the badger was kept in a tunnel. The den was about 60 square centimeters, but covered only with wire mesh on top, so that the spectators could watch the whole fight underground. A passage about 3 meters long led to the badger's den, which could also be looked into from above through a wire-mesh covering. The dog rushed into the tunnel toward the badger's lair. There ensued a desperate fight on both sides. The dog was able to pull the badger about halfway up the tunnel, but then the badger suddenly recovered its footing and pulled the dog back down to the square box. The fight now continued with twice the ferocity as before. Finally the dog's owner feared for his dog's life. They opened the wire mesh and were able to pull the dog out by the tail. The dog and badger were biting each other so mercilessly that they could be separated only with iron levers. Both animals suffered terrible injuries.

Town — Bulldogs and Badger, print circa 1800.

III. THE PURPOSES OF FIGHTING DOGS

Drawing the badger soon became a very popular sideshow in the pit. It provided a new opportunity to win or lose money by betting. Drawing the badger thus became a permanent part of the fights in the pit, along with bear baiting, dog fighting, and rat killing. Swaint reports that particularly in Scotland this sport was also staged outside the pit in the cellars of taverns, as an interesting attraction for the guests. Within Edinburgh's Royal Mile, badgers were set on by dogs in numerous establishments for the pleasure of the guests.

A Mr. Hugh Miller describes such a scene *in one of these repulsive holes, where the riffraff of Edinburgh entertained themselves with the torturing of the badger. We came through a very narrow passage into a room with a low ceiling. This was located in the center of the building, where the gleam of daylight never penetrated, and was badly lit... In the middle of this peculiar room was a trapdoor, which now happened to be open. A shrill combination of different noises rose up from the lower rooms, whereby the sharp barking of a dog and several fervent human voices — no doubt encouraging the dog — could be distinguished. At that time it was customary in gin shops to keep badgers in cramped tunnels in the cellar. Workers brought the dogs, and in such places it was a widespread sport to try out these dogs on drawing the badgers from their dens . . .*

A group of masons, accompanied by others of their class, came in. They had brought a dog, an extraordinarily unfriendly looking beast, which stared at the ground, as if it had a very bad conscience. These men were extremely interested in testing the qualities that would be demonstrated underground of their dog on the badger in the house. The tavern keeper was summoned, but he explained to the new guests that the party already down below in the cellar had come first, and that their dog was now fighting the badger. He also declared that he always had to give the badger an hour's rest before another dog could be matched against it. It goes without saying that these hours of rest for the badger always meant hearty drinking by the guests for the tavern keeper and thereby always brought grain to Boniface's mill.

Some time later the party descended through the trapdoor down a ladder to a cellar with bare walls, dark and damp, in which the stuffy air smelled like a tomb. Accompanied by loud cries and barking, the badger was again dragged from its tunnel, the blood from the previous fight still fresh on its pelt. After that the party returned highly satisfied to the tavern above and celebrated properly the courage of their dog.

The acquisition of the badger necessary for this sport already lead to much cruelty to this animal, which is so harmless and beneficial to man.

There was a great demand for badgers. Various terrier breeds were used to corner the badger on its excursions above ground. We can find a number of illustrations of this in old prints. I present first Henry Alken's *Running a Badger to Bay*. This print shows the cornered badger in defensive posture, threatened by three dogs. Now it is the hunters' task to hold the cornered badger by the neck with their nooses and to capture it alive and, if possible, uninjured. The second illustration by Samuel Alken shows badger catching with the dog and shovel.

We also present a superb print by the famous English landscape painter, Town. This shows particularly impressively two powerful, huge Bulldogs attacking the brave badger. I must emphasize here, however, that as a rule not the relatively ponderous and large Bulldog, but the Bull and Terrier was used for badger hunting, as all the illustrations from about 1820 by Henry Alken and S. Alken clearly demonstrate.

Moreover, it is also reported that the captured badgers were often pitted against

III. THE PURPOSES OF FIGHTING DOGS

individual dogs above ground, because this fight naturally was even more interesting for the spectators. Such a fight is shown in an illustration by Henry Alken, taken from his sketch book. Such fights, however, almost always ended in the badger's certain death, so drawing the badger no doubt proved to be more economical, because it spared the badger, country you even needed an official, two-legged rat catcher, who often could not catch enough rats to meet the demand of all the small pits. In Old London, rat catching was an honorable and quite profitable profession.

Specialists for rat killing in England naturally were dogs with terrier blood

Drawing the Badger: Henry Alken, London 1820.

you could charge an entrance fee, and you could bet on the drawing times.

7. FIGHTS AGAINST RATS

One of the oldest duties of our domestic dog is rat killing, because rats are followers of civilization and dangerous carriers of disease. Rat killers certainly exist in every country in the world where these vermin threaten man. The dog is an important helper of man against these pests. But leave it to "Merry Old England" to turn the gloomy handiwork of the rat catcher into a gory spectacle in the pit. In this in their veins. Fitz Barnard sees rat killing as a very special art. For this purpose you needed not only a brave dog, which did not shy away from rat bites, but one of especially great speed. The true *rat killer* bites once, drops the rat, bites again, drops again, and so on. The purebred terrier shakes the caught rat to prevent it from biting. A dog that wastes its time with shaking, however, has absolutely no chance of winning a rat killing contest, which is decided chiefly on time. The true fighting dog, on the other hand, does not shake the rat, because once it is fighting it does not

III. THE PURPOSES OF FIGHTING DOGS

Badger Baiting: Henry Alken, London 1824.

feel bites. The successful rat killer seizes the rat between head and shoulder, which prevents the rat from biting, and kills the rat immediately with a sharp bite.

Wentworth Day, a follower of this sport, sees these fights as absolutely justifiable in an ethical sense. *The rat is the most cold-blooded, savage, destructive, and disease-spreading animal in the British Isles. But you must also recognize without qualification that it is also a real fighter! And it is a pleasure to observe how a good dog kills rats quickly and skillfully. I say this not with a feeling of sadism, but as a man who admires and takes pleasure in a true fight, such as how we can love a boxing match or steer a boat against a heavy sea, swim against the pull of a strong current, or ride a horse over a difficult obstacle.*

The reality, however, could not have been quite that romantic. The same author describes one of the last old rat pits in London at the beginning of the twentieth century. *This was a rather dirty, small place, in the middle of the Cambridge Circus in London. You went down a rotten wooden stair and entered a large, underground cellar, which was created by combining the cellars of two houses. The cellar was full of smoke, the stench of rats, dogs, and dirty human beings as well. The stale smell of flat beer was almost overpowering. Gas lights illuminated the center of the cellar, a ring enclosed by wood barriers, similar to a small Roman circus arena, and wooden bleachers, arranged one over the other, rose stepwise above it nearly to the ceiling. This was the pit for dog fights, cock fights, and rat killing. A hundred rats were put in it, large wagers went back and forth on whose dog could kill the most rats within a minute. The dogs "worked" in exemplary fashion, a grip, a toss — and it was all over for the rat. With especially skillful dogs, two dead rats flew through the air at the same time...*

III. THE PURPOSES OF FIGHTING DOGS

A precise description of the betting rules for rat killing:

Basically there was a kind of weight handicap for each dog. Within a specific preset time the competing dog had to kill as many rats as the number of pounds it weighed. Here too, as with drawing the badger, we find a referee and timekeeper. For rat fights the pit was either covered above with wire mesh or additional security devices were installed on the walls to prevent the rats from escaping.

The prescribed number of rats was released and the dog was put in the ring. The clock started running the moment the dog touched the ground. When the dog seized the last rat, his owner grabbed it and the clock stopped. Rats that were thought still to be alive were laid out on the table in a circle before the referee. The referee then struck the animals three times on the tail with a stick. If a rat managed to crawl out of the circle, it was considered to be alive. The dog had to go back in the ring with these rats and kill them. The new time was added to the original time. The quickest time, the number of rats, and the dog's weight decided the victory. A rate of five seconds per rat killed was considered quite good; fifteen rats in a minute was an excellent result. Imagine catching, lifting, biting to death, and dropping a rat within four seconds . . . and seizing the next while the first is falling to the ground. We also must not overlook that the cornered rats

Badger Catching: Henry Alken, London 1820.

themselves will attack and can deliver a very painful bite. There are not too many dogs that would be willing to expose themselves to such bites!

Billy, one of the most famous rat catchers of the Bull and Terrier type, is immortalized in its record match in an old print from the year 1823. Of it we learn that *Billy is now in the possession of Mr. Charles Dew and was bred by the famous breeder James Yardington. His father is Old Billy from the kennel of John Tattersal Esq. from Woollen under Edge, Gloucestershire, and is descended from the best line of all English Bulldogs. On the maternal side, Billy is descended from Yardington's Sal. Sal comes from the Curley line. She is descended from a particularly good half-blooded bitch, well known to all sportsmen. Her father is the Bulldog Turpin, bred by J. Barclay Esq. from the Jacklin line. Barclay is the largest breeder of Bulldogs in the whole world. Turpin is a son of Blind Turk, a very famous Bulldog. The pedigree of all these dogs can be traced back more than forty years, and there are numerous old accounts about them.*

Billy killed a lot of rats before he became known in the sports world, and he won the following bets without the slightest difficulty:

1820 — Competing only against the clock, kills 20 rats in 2 minutes, 3 seconds.

1820 — Competing against Mr. Gill's Jack, kills 20 rats in 2 minutes, 8 seconds.

1820 — Competing against Mr. Germain's White Terrier, kills 20 rats in 1 minute, 45 seconds.

1820 — Competing against Mr. Baker's Tulip, kills 20 rats in 1 minute, 10 seconds.

1820 — Competing against the Walworth Dog, kills 20 rats in 1 minute, 11 seconds.

Billy — The Rat Killer of unrivaled Fame Taken from Life, Westminster-Pit, G. Hunt.

III. THE PURPOSES OF FIGHTING DOGS

Billy, the celebrated *Rat Killing Dog* (detail), London, 1823.

1822, September 3 — Competing against the clock, kills 100 rats in 8 minutes, 45 seconds.

1822, October 22 — Competing against the clock, kills 100 rats in 8 minutes, 17 seconds.

1822, November 12 — Competing against the clock, kills 100 rats in 6 minutes, 28 seconds.

1823, April 22 — Competing against the clock, kills 100 rats in 5 minutes, 30 seconds.

1823, August 5 — Competing against the clock, kills 120 rats in 8 minutes, 20 seconds.

Billy was not primarily terrier, as you would suspect at first from its appearance, but predominately a Bulldog of those times. The breed of the Bull and Terrier, the infusion of the famous shot of terrier blood, we find in its mother Sal, which for her part had a famous "half-blooded mother." Of course, these pictures again show how much the Bulldog at the beginning of the nineteenth century differed from the modern Bulldog — so much Bulldog blood and yet the anatomy of an elegant and streamlined dog!

The degree of care used in breeding these four-legged fighters is demonstrated by Billy's clear pedigree. Let us remember that the breeding of such dogs was a marvelous business if you were able to produce such a successful dog. Successful breeders were highly regarded and famous throughout the land.

The proud fighting history of Billy shows clearly the buildup of such a dog over a period of years, first with smaller fights, then serious contests against other dogs, and finally only competition against the clock! In the case of the early contests, the chronicler remembered only the year of the success, not the exact date. In the year 1822, however, began the big competitions — against the clock — with a hundred rats. Now all the dates are precisely recorded. The career was crowned on April 22, 1823, with the ninth competition. This is the one that is immortalized in the print in honor of Billy. On this day a world record was set with a hundred rats in five-and-a-half minutes. This record stood until the year 1862. Imagine killing a rat every three-and-a-half seconds constantly over a period of more than five minutes!

A glance at the print shows us that the spectators of Billy's world record were a rather illustrious group, festively dressed. This offers clear proof that in the heyday of the fights the English middle class absolutely embraced this sport as well. What a contrast to our previous descriptions from the time of the illegal fights at the beginning of the twentieth century!

According to the *Sporting Chronicle Annual*, the world record in rat killing is held by the Bull Terrier Jacko. Its contests were stopped with the following times:

August 8, 1861 — 25 rats in 1 minute, 28 seconds
July 29, 1862 — 60 rats in 2 minutes, 42 seconds
May 1, 1862 — 100 rats in 5 minutes, 28 seconds
June 10, 1862 — 200 rats in 14 minutes, 37 seconds
May 1, 1862 — 1000 rats in less than 100 minutes

There is something dry and boring about numbers; you have forgotten them before you finish reading them. Let us imagine, however, a small, bloodthirsty bundle of energy, in the middle of the rat pit for a hundred minutes — that is one hour and forty minutes — constantly killing one rat after another, a rat every six seconds! If we rearrange our statistics according to the "average speed of killing a rat," we come up with the following:

August 20, 1861 — a dead rat every 3.5 seconds, a total of 25
July 29, 1862 — a dead rat every 2.7 seconds, a total of 60
May 1, 1862 — a dead rat every 3.3 seconds, a total of 100
June 10, 1862 — a dead rat every 4.4 seconds, a total of 200
May 1, 1862 — a dead rat every 6.0 seconds, a total of 1000

Jacko set two world records here: on August 29, 1862 with an average killing time of 2.7 seconds per rat and on May 1, 1862 with his fight with a hundred rats, where it worked two seconds faster than Billy. In this connection it is certainly also interesting that this dog achieved its best speed in the killing of sixty rats, after which the average times slowly increased. The fight of May 1, 1862 is certainly phenomenal with the killing of a thousand rats in less than a hundred minutes. It is unbelievable what stamina these animals displayed in destroying the rats they hated.

According to Ash, Jacko was a black dog with tan markings of the Bull Terrier type weighing about thirteen pounds. It was owned by Jimmy Shaw in London and — you can be sure of this — this dog put food on its master's table!

There is also an account of the fight of another Bull Terrier in March 1865 owned by a Billy Shaw. On that occasion, *Pincher*, weighing about

III. THE PURPOSES OF FIGHTING DOGS

Rat Hunting: Engraving by Hancock, London, September 1, 1837.

twenty-five pounds, killed five hundred rats in thirty-six minutes and twenty-six and a half seconds, which is 4.4 seconds per rat. Jacko was able to maintain this speed only for two hundred rats, although Jacko also weighed only half as much as Pincher. According to the sportsmen in those days, this would have been a considerable handicap for Jacko.

To conclude our presentation of this sport, here is an excerpt from the *Sporting Magazine* 1825:

Rat-Killing — Billy and the Bitch from Kent

On Tuesday evening, the 10th of May, 1825, a competition was held between the quite famous Billy and the bitch from Kent. The match took place in the Westminster-Pit, which was so crowded and hot that it was necessary to remove the tiles from the roof to let in fresh air. His Grace, the King's rat catcher, appeared with a large cage full of rats of all sizes. He immediately began to put a number of them in the pit. Then the bitch from Kent was presented. She was somewhat bigger than Billy, approximately the same color, nearly pure white. The bitch now leapt from corner to corner. She did her work very well, but she apparently was not nearly as fast as Billy. The odds from the start stood at 2:1 against her, but no one accepted this bet. During the fight her mouth was washed out once, and after all the rats were dead she was removed. The timekeeper announced that she had completed her task in eight minutes and forty-five seconds. Sixty-five dead rats were counted . . .

The promoter now proposed that Billy should be given the same number of rats.

This would be the best way to assess the performance of this dog. This was accepted as fair. The pit was cleared for Billy and a fresh quantity of rats was carried in. These seemed to be even stronger, larger, and much livelier than the first ones. There obviously were also many more of them. The promoter, however, said that all of them should be put in the pit.

Billy's introduction was greeted with great joy by all. The watches were set and the task of destruction began. Billy proved to be a smooth customer, and the speed of his work was admired by all. His mouth, too, was washed out, and he went back to work until not a single rat was still alive. The work of counting began and ninety dead rats were found. A comparison of the times revealed that Billy had fulfilled this task in the brief time of seven minutes and thirty seconds. This meant that he needed one minute, fifteen seconds less time than the bitch from Kent, even though he had killed ninety rats, and she only sixty-five.

8. OTHER FIGHTS — AGAINST APES AND MONKEYS, OPOSSUMS, PIGS, HORSES, AND DONKEYS

Fighting dogs were repeatedly presented with new opponents, including those of a very unusual nature. This resulted in rather strange fights, in part with completely surprising outcomes.

Particularly interesting were the fights against the monkey, an animal that in numerous fights, thanks to its intellect and very unorthodox fighting style in comparison to the dog, achieved very surprising results.

We take the following account from *Sporting Magazine* of the year 1799: *A quite unusual fight between two animals was staged in Worcester. The wager stood at three guineas, according to which the dog would kill the monkey in at most six minutes. The dog's owner agreed that the monkey would be allowed to defend itself with a stick about a foot long.*

Hundreds of spectators gathered to witness this fight, and the odds stood at eight, nine, and even ten to one in favor of the dog, which could scarcely be subdued before the fight. The monkey's owner took a stick, about twelve inches long, from his coat pocket, tossed it to the monkey and said, 'Now Jack, pay attention, defend yourself against the dog!' The butcher cried, 'Now, get after the monkey!' He let the dog go and it sprang at the monkey like a tiger. The monkey was amazingly nimble, jumped about three feet high in the air, and when it came down landed directly on the dog's back, bit firmly in the dog's neck, grabbed his opponent's left ear with his hand, thereby preventing the dog from turning his head to bite him. In this totally surprising situation the monkey now began to work over the dog's head with his club, and he pounded so forcefully and relentlessly on the dog's skull that the poor creature cried out loudly. In short, the skull was soon cracked and the dead dog was carried from the ring. Yet, the monkey was only of medium size. This account inspired the famous English animal painter, Samuel Howitt, to illustrate this account in *Battle of the Bulldog and the Monkey*, which preserved this fight for future generations.

Another monkey we know about is Jacco Macacco. The following is an account by Pierce Egan from the Westminster-Pit in the year 1820: *The dog pit was packed in a few minutes and many people were turned away grumbling, as if they had been deprived of the most beautiful sight in the world. They were so disappointed that they*

III. THE PURPOSES OF FIGHTING DOGS

could not secure places ahead of time. Jacco Macacco was now presented in a pretty, small cage, and was greeted by the shouts and whistles of the spectators. He was not even polite enough to bow in thanks for these signs of approval, which were directed at him alone. Jacco had a the dog reached him, ducked low, with a dexterity that would serve a prize boxer well, and rolled into a ball in order to withstand the force of the collision with the dog. Nonetheless, the dog immediately dug under him and turned him over. At that moment, however, the

Battle of the Bulldog and Monkey, engraving by Samuel Howitt, published August 1, 1799, *Months Magazine* (detail).

thin chain around his waist, about two meters long, which was fastened to a steel spike, which was pounded deep in the ground. Then he was taken from his cage.

Immediately after that the dog was brought out and it charged directly at the monkey. The monkey, however, before monkey's teeth cut like a saw into the dog's throat and, like a knife, ripped a large wound.

Because of the great loss of blood, which all dogs that fought against Jacco Macacco suffered, most died shortly afterward. The monkey very rarely suffered even slight wounds in these fights. It was said of him that he was of such an unbelievably ferocious nature that it seemed expedient to his master to always have a steel plate between him and the monkey in the event that he inadvertently bit at his legs.

'What a monster!' said a greasy butcher, who sat there with open mouth, a red nightcap on his head, pointing at and resembled closely the inmates of a mental hospital, who had escaped from their straight jackets.

So much for Pierce Egan, who in all of his descriptions of the fights knew how to embellish the accounts with the reactions of the spectators. We find an illustration of this fight in Pierce Egan's book. In his sketch from

Example X: Fight with the Ape, Sir Edwin Landseer, circa 1820.

Jacco Macacco. 'I bet a leg of mutton on the monkey! You could strike me down if I ever saw such a thing before in my life. It is truly astounding! And how he punishes his opponents! He seems to destroy the dogs with such ease as if for decades he had done nothing but fight dogs!' You could fill a small book with similar quotations, which came from the noisy and excited crowd, all of whom admired the 'finishing qualities' of *Jacco Macacco*. Some laughed, others yelled wildly, and a few of the people constantly jumped up and down in a kind of ecstasy, pounded their canes on the floor, the year 1825, T. Landseer depicts a fight between Jacco Macacco and the famous bitch Puss in masterly fashion. From the year 1823 there is a print based on an illustration by S. Alken of the same monkey. Of other fighting monkeys it is reported that they specialized in blinding the dog.

With these descriptions of monkeys versus fighting dogs, we see that apparently the higher intellect combined with a healthy instinct and fighting experience made the monkey into an extraordinarily dangerous opponent for the fighting dog. It is

III. THE PURPOSES OF FIGHTING DOGS

always amazing, however, how many two-legged fools there must have been, who sent their brave dogs to certain death and thereby ensured that a spectacle of that kind could be offered to their fellow man.

Let us remain with Pierce Egan. We are indebted to his *Book of Sports* for

a fox (and therefore is also called the vulpine opossum). On the front feet it has five toes, the inner one located high up and turned inward. The toes are equipped with very powerful claws. On the hind feet we find four toes and a kind of thumb. The animal's legs are short, but powerful.

The Westminster-Pit: A Turn-up between a Dog and Jacco Macacco, the Fighting Monkey, from an illustration by Samuel Alken.

the description of the fight between an opossum and a fighting dog in London, a rather peculiar story — marsupial against fighting dog! *Mr. Ferguson had a young Terrier bitch, about 16 months old, liver colored, weighing about twenty-five pounds. Mr. Jenkins owned an opossum, which he had brought back from New South Wales. He thought it was about three years old and that it weighed about twenty-six pounds. An opossum is similar to but smaller than*

The bitch and the opossum fought on January 6, 1829. It was a raw day. The fight had to be moved from Humpton Green to an old barn, because so many wanted to see this fight at any price, such was the degree of interest in the whole community. Many bets were made before the fight. In so doing guineas were laid against pounds. Possey, the opossum, was the favorite of our Norfoth experts, who knew well the nature of this wild animal and had seen how superbly the

III. THE PURPOSES OF FIGHTING DOGS

trainer Jimmy Neal had prepared the opossum. They drove the odds up to 3:2. I heard that even 2:1 was offered, but I cannot confirm this. The bitch was trained by Tom Riffley.

Round 1. Possey came out very fit, shook his bushy tail, and lunged at the bitch like a flash, bit her on the shoulder and tore out a piece of flesh. Then he retreated, jumped again at the front legs of the bitch, but missed. In the meantime the bitch was not inactive and made several attempts to get a firm hold, but the gentleman's long fur fooled the poor bitch, who tore out a mouthful of the outer coat each time she bit. Finally she caught him, 'where the Irish-men put their lundy,' bit him very badly, while Possey took revenge and used his claws to scratch the bitch horribly. Finally he freed himself and was taken to his corner.

Round 2 began after two minutes. Both charged at the same time, cracked their heads together, and fell down. Possey turned around and seized the bitch by the throat, threw her over him, stretched the hind legs straight out and choked almost all the air out of the bitch. Odds of 4:1 were now offered on Possey, but no one took the bet. The bitch fought in subdued fashion until she got her wind back. Then she attacked again, seized him by his long snout, and dragged him in good style through the ring, despite his

Tom and Jerry sporting their Blunt on the phenomenon monkey Jacco Macacco at the Westminster-Pit. Print by L. R. and C. Cruikshank, London circa 1820.

III. THE PURPOSES OF FIGHTING DOGS   117

claws, which ripped frightening wounds in the bitch. Possey escaped again and was carried to his corner. This round lasted nine-and-a-half minutes.

Round 3. *The bitch attacked first and seized Mr. Possey by the snout, held tight and dragged him through the ring for about two-and-a-half minutes. The*

Then she seized him by the shoulder, got a superb hold, and now for the first time Possey screeched loudly. When he got free he went to his corner. It proved impossible to convince him to continue the fight. Accordingly, the dog was declared the winner. The whole fight lasted thirty-seven minutes.

Running a Badger to Bay: Henry Alken, London circa 1820.

opossum fought only with his claws. When she lost her hold she sprang at the back of his neck, which she had cleaned of fur in the previous rounds. She worked him over vigorously, shook him back and forth, until he got loose and was again taken to his corner. Possey had become quite weak because of the loss of blood, but was made fit again by Riffley, who rubbed something on his nostrils.

Round 4 and the conclusion. *The bitch again seized the foreigner by the nape and left behind clear tooth marks.*

Now to another peculiar fight. This took place on March 18, 1849. We know little about this fight, but we do have an illustration and some information on the weight of the opponents and the wagers. Because the wagers are expressed in dollars, it seems likely that this fight took place in the United States.

One of the opponents was *the famous fighting pig, Pape,* which was probably a wild pig of very small size. Its fighting weight was given as 34 pounds. Our illustration shows the dwarf size of this animal in comparison to the men and dog. Pape fought Crib, a brindled Bull and Terrier, with a fighting weight of 46 pounds. Crib, undefeated until then, was beaten after a short fight with Pape and later died from the wounds he suffered in this fight. On the same day Pape killed a second dog named Imp. This dog had a fighting weight of 17 pounds, and the fight lasted only one minute and seventeen seconds.

In his *Encyclopedia of Rural Sports*, Mr. P. Blaine reports about *horse baiting*, the fight between the dog and the horse. From this book we also have an illustration. The fight took place in the year 1682. Blaine reports: *In public notices it was announced that on April 12 a horse of unusual strength and eighteen to nineteen hands high was killed in His Majesty's Bear Garden at Hope on the Bankside in London. These festivities were staged in honor of the Ambassador from Morocco and other noblemen, who either knew the horse well or were prepared to pay the high entrance fee. It appears that this horse had originally belonged to the Earl of Rochester. This horse, of a very wild character, had killed various other horses and had been sold to the Earl of Dorchester because of these misdeeds. Here it carried out a lot of other crimes and was therefore sold to the "worst of the savages," to those who ran the Bear Garden. On the appointed day various dogs were set on the wild stallion, but they were trampled or driven out of reach of his hooves. Encouraged by these events, the owner decided to keep the horse for another day of sport, and he sent his assistants to lead the horse away. But before the horse had reached London Bridge, the spectators demanded fiercely that the promise to worry it to death be fulfilled, and they began to tear the building apart. Finally the poor animal was brought back and other dogs were set on it without success. Then it was killed with a sword!*

Ass baiting, the fight with the donkey, was quite common at that time, but apparently was never very popular. This had less to do with not

The Celebrated Fighting Pig, Pape fighting Crib on March 18, 1849.

III. THE PURPOSES OF FIGHTING DOGS

A Horse Baited with Dogs, from a manuscript, fourteenth century, Royal Library (from Strutt).

wanting to torture this animal, but rather that the poor donkey offered little appeal for the sport, because it was rarely possible to make it attack the dogs.

Surely there are also other animals that had to suffer under the "baiting fever" in Old England. We believe, however, that with the examples presented in this chapter have given our readers a quite compelling and impressive overview.

9. FIGHTS AGAINST MAN

This section is of decisive importance for understanding our fighting dogs. In previous sections we gave a broad overview of the various dangerous opponents for our fighting dogs. We must emphasize, however, that only in the discussion on war dogs did the dog fight against people. The remaining sections described fights between animals.

For good reason our fighting dogs in England were provoked to attack humans only in very rare cases. The Englishman no doubt recognized early on that fighting dogs did not make good guard or police dogs. We have seen repeatedly how ferocious these animals are. To train these large dogs to attack people is insanity!

We know that the fighting dog breeds were left entirely to their own resources through centuries of animal fights. For this they needed self-confidence, absolute courage, and the will to win, even if they had to pay with their own life. These animals lack — intentionally, from centuries of breeding — the willingness to submit to any other will in the fight. In these fights the submission to the opponent's will would have been fatal to the fighting dog.

We cannot break down character — whether in human beings or animals

III. THE PURPOSES OF FIGHTING DOGS

Donkey attacked by Staffords, oil painting circa 1840.

Gentleman and the Bull Dog. The *Sporting Magazine*, 1801.

III. THE PURPOSES OF FIGHTING DOGS

— as we see fit. This is also impossible to do in breeding. The insensitivity to pain when enraged, the willingness to fight to the death, these cannot simply be left out on human command. These dogs fight, and — once they are fully enraged — they can scarcely be controlled by man until the end of the fight. In the discussions of the animal fights, we have seen that the dog's jaws usually must be forced open to separate it from its victim.

It is unfortunately a widespread fallacy that fighting dogs, because of their constant willingness to fight, would be particularly suitable for training as police or guard dogs. When you are aware of the size, the weight, and the unflagging energy of our fighting dogs, along with the readiness to fight mercilessly, you will realize that these dogs were consciously used by our ancestors only for fighting other animals, and — almost never — for fighting man.

I have included the following two examples, where fighting dogs fought against human beings, in this book for the purpose of warning against doing the same thing with your own dog, and because they simply serve to round out the book. Whoever sees this as a warning against setting large fighting dog breeds on people has understood the author's intentions.

In *Sporting Magazine*, volume XVIII, there is the description of a fight between the *Gentleman and the Bull Dog*. The *Sporting Times* reports on this fight, which took place in 1801:

A fight between a man and a Bull Dog took place some time ago to settle a bet. With its first charge the Bull Dog already succeeded in throwing and pinning its opponent. Although the dog's jaws were nearly closed by a muzzle, it succeeded in sinking its teeth into the man's body. Had the dog not been pulled away immediately, it would have disemboweled the man. The aforementioned illustration shows a large, mastiff-like, white dog charging its opponent. In this fight, despite the handicap of the muzzle, the dog was the winner.

Now to another account. A journalist named James Greenwood reported in the *Daily Telegraph* on July 6, 1874 about a fight between a man and a Bulldog, to which he had accidentally been eyewitness. This fight took place on June 24, 1874 in an old inn in Hanley, Staffordshire. At this time in England there were strict laws banning all animal fights. For that reason, Mr. Greenwood's account in the *Daily Telegraph* prompted a great outcry and intensive investigation. Precisely for this reason, however, the journalist's account was also largely substantiated.

Now to the fight in Hanley. In a large guest room a ring was cordoned off with a line. The spectators, who eagerly awaited the third fight between Brummy, a bow-legged dwarf, and Physic, a stately, white Bulldog, were mostly coal miners from the Black Country. Among them, however, there were also gentlemen from better social classes. All in all there were about fifty spectators. The floor was covered with sawdust, from the ceiling hung an oil lamp, and all the windows were closed and carefully covered. The only ventilation was through the fireplace. Thick smoke from cigars and pipes as well as the perspiration from the crowd made the room hot and sticky.

The dwarf Brummy, about forty years old, was at most four-and-a-half feet tall. The enormous size of his head and ears were particularly striking. He had huge hands and feet and was extremely bow-legged. Let us now follow our eyewitness account:

Brummy took off his coat, his jacket, his blue shirt, and his shoes, and stood there only in his pants and a dirty,

Alfred Concanen: Physic and Brummy, An Evening at Hanley.

armless undershirt. Half-dressed in this way, you could see that his build was extraordinarily muscular. His arms were covered with scars and hair. He carefully oiled his whole body, after the oil had previously been checked by the referee and the rest of the court of honor.

'Everything in order?' asked the referee. 'Yes, bring him in, whenever you want!' grinned the dwarf, and you could hear the scratching of a four-legged animal, an eager whining, and through the open kitchen door came a frightful, dirty white Bull Dog. As soon as Physic saw Brummy, it growled ferociously, but it was held back by a wide leather collar and tied to a heavy chain anchored in the wall. Another assistant fastened a strong line to the dwarf's belt. Each of the two fighters had enough room to move that they could meet in the center of the ring. Now there could no longer be any doubt about the ensuing frightening duel. The horrible dwarf himself had placed a wager on this fight against Dan'l's Bull Dog, and his friends had also placed large bets. A spectator, after taking a healthy swig from the rum bottle, said, "It has been a draw so far. This is their third fight. This time there will be a decision!"

The rules were set up such that the dog and the man had enough rope for them to meet in the center of the ring. The rope was short enough, however, that each could avoid the other by retreating when it seemed the smart thing to do. The two-legged beast could kneel or move on all fours, which he seemed to prefer. His only weapons were his fists. The dwarf was not allowed to grab the dog's wide

III. THE PURPOSES OF FIGHTING DOGS

collar. He was only permitted to attack the dog when the dog attacked him, and he could only defend himself with his hands. If the Bull Dog succeeded in getting a proper hold on its opponent, the man need only shout, "I'm done!" and immediately measures would be taken to force the dog to loosen its grip. It was Brummy's task to punish the dog, to stun it or to beat it, so that, despite the encouragement of its master, after a pause of one minute it would be unable to continue the fight against the dwarf.

Dan'l, the owner of the Bull Dog, had a bucket with water and vinegar and a sponge, which he could use to revive the dog if necessary. One of the dwarf's friends handed him a big bottle of brandy, from which he took a long pull and then set the bottle within reach in his corner.

Now the man pulled his flannel undershirt down from his throat, spit in his enormous hands, clenched his fists — as big as sledgehammers — and kneeled down grinning. In the meantime Dan'l had finished with his prefight preparations of Physic. Man and dog stood ready to fight.

The red-eyed Physic needed no encouragement; it was only too eager to begin the fight. It did not bark, but it was so filled with rage that saliva dripped from its blunt snout and its panting became increasingly louder and more hysterical. It truly required no further encouragement before the first round. As soon as the referee cried, 'Let go!' this dirty, eager beast leapt forward with such ferocity that the chain gave a loud clang as the individual links snapped against one another.

The dwarf, however, was by no means knocked down and defeated by the first onslaught. Once this gruesome fight began, there was a terrible tension in the ring. The man crouched on all fours on the ground at the command, 'Let go!' and he calculated exactly the length of the chain that tethered his foe. Like a cat the dwarf leaned back, just beyond the reach of the fangs, and then smashed his two fists in the middle of the dog's skull. This nearly brought the dog to its knees, but it recovered very quickly. Before the dwarf could retreat, Physic rushed forward a second time, and this time the teeth seized the dwarf's arm and left a shallow red furrow. The dwarf grinned in wrath and sucked on the wound. There was great excitement among those who had bet on the Bull Dog. They clapped their hands in joy and took delight in the first sight of blood.

The hairy dwarf still grinned as Dan'l held his dog and readied it for round II. The dwarf did everything to provoke the dog even more and showed it his bleeding arm. The animal, perhaps blinded by the initial success, charged suddenly and leapt at its foe, but this time the dwarf met the Bull Dog and gave the dog such a smashing blow under the ear that it was knocked down. At that moment the dog apparently was stunned and now started to bleed heavily, to the frenetic joy of the friends of the human beast. But they soon became serious again, for with amazing energy Physic turned around and again leapt at the dwarf. This time, however, it managed to sink its teeth in the hairy arm, leaving a horribly deep wound when the dwarf yanked his arm from between the ravening jaws. The Bull Dog licked its lips and had fewer tears in its eyes than its master, who proudly carried it back to its corner. The dwarf went to his corner for a drink of brandy and to wipe himself off with a towel.

He was soon ready and grinned again as round III began. This time the fight raged in all its fury, the dog sinking its teeth into the man, the man raining terrible blows of his sledgehammer-like fists on the dog's ribs and head, until finally the man's arms were covered with blood. Behind the ropes this gruesome business was rated as 2:1 in favor of Physic.

To sum up the next seven rounds, in their details they were so appalling that more than once I would gladly have left my place had I been able to do so. The rest of the crowd, however, was made of far sterner stuff. The more ferocious the gruesome fight became, the more they liked it, and in their excitement they leaned over one another and the rope, and screamed and snarled guttural noises, when a good blow or bite met their fancy.

When round X arrived, it was apparent that the Bull Dog's skull had swollen to far more than normal size. It had lost two teeth and one of its eyes was swollen completely shut. The dwarf's fists and arms, on the other hand, were covered with blood, his terrible face was deathly pale from his exertions and because of doubt about his victory. Fate, however, was kind to him. In round XI the Bull Dog charged again, fresh and flashing with rage and with frightening persistence. With the strength of desperation the dwarf landed a terrific blow under the chin. The dog was thrown with such force against the wall, that despite all its owner's efforts it remained lying for more than a minute. The monster, however, who had desecrated all that was human in him in this fight, was declared the victor.

James Greenwood recorded this horrible experience in his book, *Low Life Deeps*, as "*An Evening at Hanley.*" The illustration by Alfred Concanen gives us a striking picture of this gruesome and degrading fight.

10. THE BANNING OF ANIMAL FIGHTS

Caricature of a treadmill.

The previous section showed us an England, whose inhabitants found a diversion from the gray daily round, who got carried away by the courage, the fearlessness in the face of death, the fight to the bitter end. Psychology no doubt has an explanation for why it is precisely the man who is constantly humbled and abased by his environment who dreams the dream of his own heroism, but who cannot realize it for himself because of the social structure in which he lives. What would be more obvious than vicarious experience, the acquisition of your own fighting dog? What man himself cannot achieve, his dog has the chance to accomplish, if only it has the right combination of courage, insensitivity to pain, strength, stamina, and skill. To this was added the chance at relative wealth. Success of the dog and skill in betting brought winnings to the lower class, which were enormously high compared to the earnings from the daily labor. It is also understandable that the sportsman infected with fighting and

III. THE PURPOSES OF FIGHTING DOGS 125

betting fever always saw only the successes, not the high losses that inevitably accompanied them.

At the beginning of the nineteenth century there arose a counter movement from the circle of those who saw in the animal a living creature entrusted to mankind, which knew pain and suffering just as man did. It must have been very hard, however, for these animal lovers even to have been heard by the sports-blinded masses. In the year 1777 we find the first local attempt to make bull and bear baiting a punishable offense. In Wolverhampton the city council introduced a law — unfortunately, in vain — that threatened anyone who participated in bull or bear baiting with a fine of five pounds. In the House of Commons in London, in 1802 a bill was introduced that would ban all animal fights in the whole country, but it was defeated. In the public opinion of the nation this was considered to be a plot

Terrier at the Fox Den. This bronze by P. J. Méne from the middle of the nineteenth century shows two Bull Terriers with a Cairn Terrier at the fox den.

hatched by the Jacobins and the Methodists with the goal of making life dull and supporting the government. The bill was defeated with a majority of thirteen votes.

Nevertheless, the speech of Representative Sheridan is worth retelling. We cite from it: *What sort of moral model does the farmer set for his wife and children? He sells his bull, so that it can be worried to death by dogs. You must witness how this poor, harmless animal is attacked by the dogs, how the bloody tongue is torn out of the mouth by the beasts. It is the same animal that was protected, cared for, and loved for years. But the cruelty toward the bull is not the only thing to deplore in these fights. What kind of example does the farmer set for his children, when he takes his old Bull bitch, for many years a faithful guardian of house and courtyard, together with her pups to the bull ring, in order to prove the bravery of his dogs? He brings the bitch into the arena and sets her on the bull that is foaming with rage. She seizes the bull by the snout and forces it to the ground. But what is the master's reward for his favorite, amidst the cries of jubilation of the crowd. He picks up hedge clippers and — to prove the keenness of his dog — he dismembers the bitch with the clippers while she still hangs onto the bull! Finally, the farmer sells the bitch's offspring for five guineas apiece!*

Is it not extraordinarily interesting that here in the English House of Commons in the year 1802 we again encounter the already well-known story of the fighting dog being dismembered alive. Was Mr. Sheridan aware that in 220 A.D. Aelian already reported of a similar story from the campaign of Alexander the Great in India, except that here the dog's opponent was a lion?

After the defeat in Parliament, responsible animal lovers continued to fight throughout the land for the abolition of the fights. In the *Annals of Sporting for 1822* there appeared an

The Bear Garden and Hope Theatre 1616.

III. THE PURPOSES OF FIGHTING DOGS

Dogfight Vignette, sketch for a placard, water color by the painter Rowlandson, circa 1800.

interview between the owner of a dog pit in London, Charley Eastup, who was considered an eccentric in his day, and a Mr. Martin, one of the leaders of the humanitarian movement. Mr. Martin visited the pit to get a personal impression there. Here is the outcome:

"I would like to see the animals that fight in the pit."

"With the greatest pleasure, Sir!" Charley bowed to Mr. Martin and led him into the pit.

"Pardon me, what is your name?"

"My Lord, Sir, do you not know Charley Eastup?"

"No, actually not. Now, what do you have to show me?"

"Here, Sir, this is the bear, and it is a very costly one. There is not a dog in all of Westminster that could hurt him. Here I have three badgers in their stalls, should I show them to you, Sir?"

"Yes, please."

"Here, my Lord, take a good look at this pretty fellow. He is a very valuable 'cutter,' let me tell you. He is one of those who understands how to give and how to take. You can see this in all his fights. Yes, Sir, you will hardly believe it, but on Wednesday last this badger was 'drawn' over two hundred times, and today he is as lively as a newborn eel."

"What is this 'drawing'?"

"My Lord, Sir, you don't know that? Now, you see, we put the badger in this long tunnel, at the end of which there is a door. Then I stand next to the tunnel and open the door. Now whoever happens to want to

"Beware of the Dog," bronze 1880, P. Lecourtier. Collection of the author.

III. THE PURPOSES OF FIGHTING DOGS

draw out the badger with his dog drops to his knees, holds the dog tightly by the nape with his left hand and with the right holds onto the tail. Then he lets go with the left hand and lets the dog go at the badger. The dog holds on tightly, if it is worth anything at all. Then the dog owner pulls his animal out by the tail, together with the badger. I grab the badger by the tail, the man sticks his dog's tail in his mouth and bites hard, then the dog lets go. The badger drops to the floor, I toss it back into the tunnel, and close the door again. This is what we mean by 'drawing.'"

"Now, I find this absolutely cruel. I find it incomprehensible how you can find pleasure in this!"

"My Lord, Sir, this is by no means cruel, once you are accustomed to it. I can see that you do not belong to the 'fancy,' Sir, so you also cannot properly understand these things. But I will gladly explain everything to you. You see, I run this pit with one bear, two or three badgers, and two or three 'hack-dogs.' A 'hack-dog' is one that is kept to provide real sport when there are no good dogs from the audience. I stage fights two days a week, Mondays and Wednesdays. I charge six pence for admission and let anyone bring a dog if he wants to. Only at evening performances, which are staged separately for special fights, do I collect an entrance fee of one shilling from a poor man, and a gentleman, like yourself, my Lord, he gives what he thinks is proper. Now, Sir, when the sport begins, first we pair off the dogs. Do you understand

Bear Garden.

what that is?"

"Unfortunately, no."

"Now, that means that two are always brought together who want to put together their dogs for a 'turn-up.'"

"Turn-up, what is that?"

"My Lord, Sir, I see that you are like a child! Turn-up, this is a fight, one dog against the other, what else could it be? Either 'from the scratch' or however they want. No, when they are satisfied with these things, then I take out the bear, and those who want to 'run at him.'"

"Run at the bear? I would truly be injured, if I ran at the bear." Outside they could hear the bear growling cheerfully after his midday meal. Mr. Martin goes to the door of the pit and says somewhat crossly: "Come, you take me for the fool, I would like to see the man who runs at a bear."

"Ha, ha, ha! Have no fear of him, he is as gentle as a newborn lamb. My Lord, Sir, how ignorant you are. Dogs run at the bear, not men. Ha, ha, ha, oh my Lord, I wish that Jack Goodlade and Jos were here, how they would laugh. Now Sir, I really would like to explain everything to you. The dogs go at the bear, until they have had enough, and they soon have enough, for he is an experienced bear. After that I lock up the bear again and bring out the badger tunnel, for those who would like to draw the badger two or three times. Naturally not the men, Sir, surely you understand, but dogs and bitches, for it gives them great pleasure, when they are of the right sort. Then, when all have had their chance, I shut the tunnel, blow out the lights, and everyone goes on his way."

"But you have in no way convinced me that this sport is not cruel. For naturally the dogs bite the bear and the badger terribly, and in turn the dogs are bitten."

"Now yes, Sir, they all bite, as well as they can, but they do not get hurt and if they do, then I can do nothing about it. The ape, the poor fellow! Oh sir, I have lost the best animal that ever fought in this pit. Poor Jacco! He was the fellow who bit! But last Wednesday, they took all the bite out of him. The damned cur tore off his lower jaw, and he was dead as a sheep in half an hour. Now, the dog died too, that is my solace!"

"Why, was he also bitten?"

"Yes, in fact, he was well bitten. The poor Jacco cut his windpipe in two pieces, nice and clean, and so he went to the Devil too, five minutes after my poor Jacco died in my arms!"

"Monstrous! I will bring this up in Parliament!"

"Now sir, there are only three or four pits in all of London and probably one or two bears are killed a year in them, and maybe a hundred badgers and maybe twenty or thirty dogs!"

This visit with Charley Eastup gives us a last drastic impression of the events in the pit, but at the same time also shows how naive and ignorant the honorable members of the House of Commons were of these events.

Nonetheless, in the individual counties, measures were introduced against this cruelty to animals, as we can see from a story in the newspaper the *Lichfield Mercury* in October 1828: *The cruel and repulsive scenes in Greenhil Wakes on Monday and Tuesday evening were publicly prosecuted at City Hall. As we reported last week, the principal organizers of these brutalities were not residents of the city. As we also discovered, they belong to a*

III. THE PURPOSES OF FIGHTING DOGS

Tyger and his Master. The Duke of Hamilton with a fighting Bulldog. Engraving circa 1790.

band of scoundrels who buy bulls and travel with them from town to town to let them fight dogs. They receive considerable sums of money for their inhuman activities. It was then further reported that those responsible for these fights officially sell their bulls to straw men, so as to be able to stay in the background in the event of a fine. The inquiry, however, proved the guilt of the instigators, all of whom came from Wednesbury and its surroundings. Each of them had to pay a fine of two pounds, ten shillings, and because they allegedly had no money, instead they were sentenced to a penalty of three months in prison. They were punished under the General Turnpike Act 3, George IV, c. 126, 121st section, according to which all persons are to be punished who are on or near public streets where there are organized bull fights that disturbed others.

Naturally, these penalties seemed small in comparison to the profits earned from bull baits. The jailed offenders surely knew how to buy their way out quickly. Substantially tougher laws would be needed to effect a change here.

In 1829 a second bill was introduced in the House of Commons, but it was defeated by a majority of forty-five votes. The defeat was based on the view that the poor in the land would suffer too many restrictions on their few pleasures. It would have been very unwise politically to take away this entertainment from the lower class. If someone happened to die occasionally, it was his own fault. It was emphasized that there was the danger that through such a prohibition the valuable ancient breed of the English Bulldog — the symbol of the English national character — could die out.

The supporters of the law argued on the basis of the ethical laws of humanity. They stressed that precisely this English Bulldog played a quite dubious role and had participated terribly in the evil scenes of cruelty throughout the land. For a part of the population of England, the Bulldog now became a "criminal among dog breeds," a "monster of savagery."

Following the renewed defeat, the animal lovers further mobilized the public opinion through laborious, painstaking work. The Anglican Church now also took the side of the animal lovers and supported the ban. In 1832, in the Black Country, the coal and steel center of the country, a terrible cholera epidemic broke out, which depopulated the whole countryside. Despite continuing new protests, the "bull hangers-on" decided to stage their cruel mass entertainment even in this melancholy time. Sunday was to be a day of prayer for those afflicted by the plague throughout the land. The organizers chose the same day for their cruel animal fights!

The English Parliament now could no longer ignore the weight of public opinion, despite the influence of those interested in continuing the fights. Accordingly, Parliament passed a law in 1835 that prohibited any kind of animal fight under threat of punishment. This law prohibited operating a house, a pit, or any other place for the purpose of allowing a bull, bear, dog, or any other animal to fight there.

Nonetheless, there are many reports indicating that many years would have to pass before the law could be enforced. Until then bull baiting

III. THE PURPOSES OF FIGHTING DOGS

continued in cities like Wildenhall, Tipton and Bilston. The bear and bull fights, however, were comparatively easy to detect, because, of course, more space was needed to stage them. With these fights the police were able to enforce the new law fairly soon. Dog fights, on the other hand, went underground. They may even have increased in popularity, because the other fights soon became impossible to stage. It took another seventy years or more before dog fights were eliminated in England for all practical purposes.

Nonetheless, we must keep in mind that even today, animal fights still take place. We have already proved this in the discussion of dog fighting. So, there remains only the slight hope that mankind throughout the world will no longer compromise its dignity and will no longer abuse its animals. Nothing against our fighting dogs, for they have found new tasks, as we will still see, but put an end to cruelty to animals!

III. THE PURPOSES OF FIGHTING DOGS

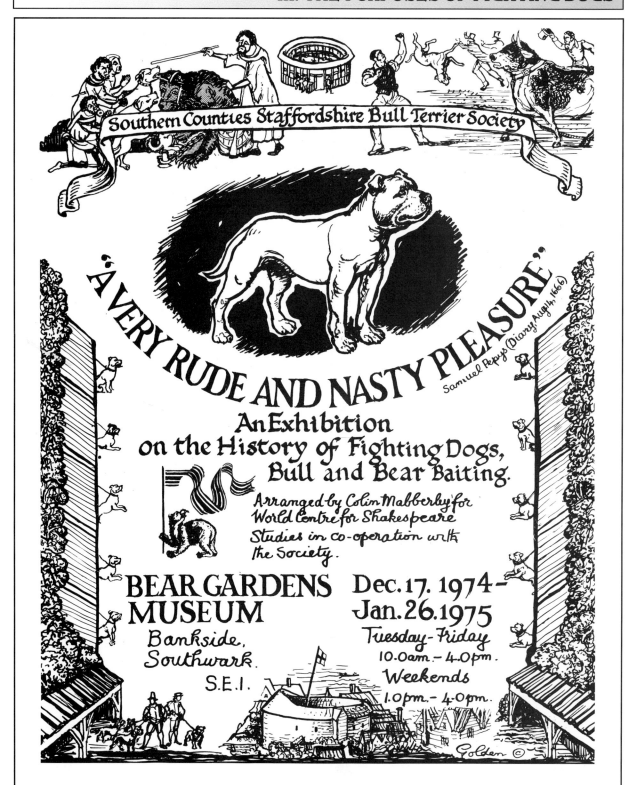

SUGGESTED READING

TS-235
The Working Pit Bull
320 pp, over 300 color photos.

TS-142
The Truth About The American Pit Bull Terrier
320 pp, over 350 color photos.

TS-141
Pit Bulls and Tenacious Guard Dogs
320 pp, over 300 color photos.

H-1063
The World of The American Pit Bull Terrier
288 pp, Color & B/W photos.

H-1069
The World of Fighting Dogs
287 pp, Color & B/W photos.

PS-613
This is The American Pit Bull Terrier
176 pp, B/W photos.

H-1024
The Book of The American Pit Bull Terrier
350 pp, B/W photos.

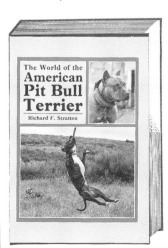

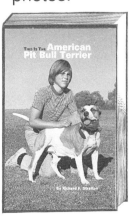

INDEX

Page numbers in **boldface** refer to illustrations.

A

Aelian, 126
Aldrovandus, 22
Alexander II, 26
Alexander the Great, 48, 50
Alken, Henry, 51, 52, 53, 54, 55, 57, 59, 63, 63, 67, 74, 75, 79, 89, 92, 96, 100, 101, 104, 105, 106, 107, 117
Alken, Samuel, 104, 114, 115
Alyattes, 20
American Bull Terrier Club, 86
American Pit Bull Terrier, 86
Ancestors of dog, 13
Anecdotes, **60**
Annals of Sporting for 1822, **126**
Ash, E. C., 13, 110
Ass baiting, 118
Attack training, 119
Auroch hunts, **25, 26, 38**
—oil painting, **73**
—porcelain 1760, **65**
—print, **27, 28**

B

Badger baiting, 99
—print 1824, **106**

Badger Catching, **107**
Bailey, Ken, 99
Baiting the Bull, **81, 84**
Ban of animal fights, 124
Ban of dog fights, 86, 99
Barclay, J., 108
Barnard, L. Fitz, 82, 105
Barsbeck (Germany), 14
Battle of Marathon, 20
Battle of the Bulldog and Monkey, **113**
Bear, 38
Bear baiting, 48
—print 1796, **52**
—print 1820, **54**
—print 1821, **49**
—print 1823, **53, 57**
Bear Baiting, **51, 52**
Bear fight, print, **40, 41, 129**
—print 1616, **126**
Bear Garden Museum, 54
Bear Gardens, 54
Bear hunt
—oil painting 1800, **58**
—print 1608, **44**
—print 1650, **44**
—print 1797, **45**
Bear Hunt, **38**

INDEX

Bear hunt in winter, **46**
Bear hunting with Dogs, **43**
Beckmann, Ludwig, **36**
Belcher, **90**
Billy, 108—110, **108, 109,** 111
Bison, 25—26, 38
Blackface, 56
Blaine, P., 100, 118
Blondus, 22—23
Boar, 38
Boar hunting, 26
—bronze, **33**
—oil painting, **56**
—plate 1600, **62**
—porcelain, **61**
—print, **68**
—woodcut, **36**
Boar hunt in the English Colonies, 35
Boar Hunt, The, **26**
Book of Sports, **115**
Bosch, 81
Brabant Bull Biter, 79, **81**
Braunschweig, Heinrich J. von, 29
Breed development, 14
Bruin, 57
Brummy, 121—124
Buffon, 15
Bull and Mastiff, **The,** 71
Bull and Terrier, 18, 39, 90, 104
Bull Attacked by Dogs, 81, **83**
Bull baiting, 60, 65, 78
—print 1799, **64**
—print 1800, **70**
—print 1817, **55**
—print 1820, **59**
—print 1823, **63**
—rules of, 65
Bull Baiting I, **74**
Bull Baiting II, **75**
Bull Broke Loose, 67
Bull-Baiting, **81**
Bull-Baiting 1791 Captain Lad, 79
Bull running, 67
Bull Terrier, 97, 100
Bulldog, 104, 108, 132
Bulldogs and Badger, **102**

C

Carl V, 21
Charlemagne, 26
Chincha Bulldog, 19
Chouking, 16
Circus, 61, 106
Classification of dog, 15, 19
Concanen, Alfred, 122, 124
Congress, 86
Crib, 91, 118
Cromwell, Oliver, 54

D

Daily Telegraph, 121
Danish Dog, 18
Day, Wentworth, 106
Dew, Charles, 108
Dog fights, 81

—basic elements, 87
—opposite sex, 90
—rules, 87
—print 1811, **88**
—print 1824, **92**
Dog Fight in the Street, **89**
Dog Fight, A, **90**
Donkey Attacked by Staffords, **120**
Donkey baiting, 118
Drawing the badger, 100
Drawing the Badger, **105**
Dutch Bull Biter, **82**
Dwarf fights, 121

E

Earl of Essex, 22
Eastup, Charlie, 127, 130
Edward the Confessor, 50
Egan, Pierce, 51, 52, 60, 112, 114, 115
Elizabeth I, 22, 52
Encyclopedia of Rural Sports, 100, 118
English Mastiff, 17, 26

F

Feddersen-Weirde (Germany), 14
Few Real Fanciers, A, 67, **60**
Fight with the Ape, **114**
Flemming, 79
French Bulldog, 18

G

Gentleman and the Bull Dog, **120**, 121
Geo, 95
George IV, 132
Grand Duke Nicholas, **72**
Great Dane, 18
Greenwood, James, 121, 124
Greyhounds, 14

H

Hackwood, Frederick W., 72, 101
Hammurabi, 20
Hancock, 111
Hellfarth, Carl, 66
Henry VIII, 21
Hondius, Abraham, 68
Horse baiting, 118
—print, **119**
Howitt, Samuel, 32, 41, 45, 58, 70, 71, 111, 112
Hughes, 91
Hungry Tiger, 98
Hunt, G., 108
Hunting dogs, 14, 25

J

Jacco Macacco, 112—114
—print 1820, **116**
Jacko (Bull Terrier), 110

INDEX

Jacobsen, Juriaen, 29, 35
James I, 51, 53
John, 61

K

Kaendler, Johann Joachim, 26
Kambyses, 20, 48
Kandler, J.J., 65
Keller, 16
Kramer, 16
Kreiger, Louis, 91

L

Landseer, Edwin, 81, 83, 84, 114
Landseer, T., 114
Laneham, 53
Las Casas, Peter, 22
Lecourtier, P., 128
Licking, 96
Linnaeus, Carl, 15
Lion baiting, 48
—print, **50**, **51**
Lloyd, Charles, 91
Lloyd, H., 20
Low Life Deeps, 124

M

Macks, Jack, 99
Marmont, 22
Martin, 127
Master Brown in distress, **42**
Mastiff, 17, 26
Match at the Badger, A, 100, **96**, **101**
Meeks, Jack, 95
Mellish, Mr., 90
Memoirs of the Pit, **95**
Menzler, H., 27, 34
Miller, Hugh, 104
Misson, 63
Mongrel, 95
Monkey baiting, 112
Montgomery, E.S., 58, 97
Moritz, Count, 29
Munchhausen, Baron von, 82

N

Napoleon, 22

O

O'Gaunt, John, 67
Old Bear Garden, 54
Old Billy, 108
Old Sal, 91
Old Storm, 90
On the Bear Hunt, 32
Oppian, A.D., 26
Opposum baiting, 115
Origin of dog, 13
Origin of fighting dogs, 15

P

Paddington strain, 90
Paddy, 97
Pape, 118
Parliament, 86, 126, 132

Philipp, Count, 28
Physic, 121
Physic and Brummy, **118**, **122**
Pig hunting, print, **35**
Pilot, 91
Pincher, 110
Police Gazette, 91
Porcupine hunting, **48**
Possey, 115
Pug, 17
Purebred Ox, The, **79**
Puritans, 54

R

Raging Boar, A, 29
Rat baiting, 105
—betting rules, 107
—records, 108, 110
Rat Hunting, **111**
Real Life in London, **52, 56**
Red Lady, 98
Ridinger, Johann Elias, 26, 27, 30, 32, 34, 46
Rowlandson, 90, 127
Running a Badger to Bay, 104, **117**

S

Sadler, Justus, 32, 44
Sanderson, George P., 39
Scaurus, 28
Shaw, Billy, 110
Sheridan, 126

Snyder, Franz, 29, 38, 40, 43
Society for the Prevention of Cruelty to Animals, 86
Sopeithes, 49
Sparks, 86
Specht, Fr., 38, 42
Sporting Chronicle Annual, 110
Sporting Magazine, 52, 71, 86, 90, 111, 112, 121
Sporting Times, 121
Stag, 38
Staverton, George, 65
Stern, 98
Stonehenge, 15
Stradanus, 29, 50, 61, 78, 79
Strebel, 16, 18
Stuart, Mary, 51
Studer, 15
Swaint, 104

T

Tantzer, Johann, 26
Tasting, 96
Tattersal, John, 108
Tempesta, Antonie, 25, 28, 44
Terrier at the Fox Den, **125**
Tibetan dog, 16
Tibetan Wolf, 16
Tiger, 59
Town, 102
Training, 96
Treadmill, 95

INDEX

Trix, 97
Turpin, 108
Tyger and his Master, **131**

U

Ueka, H.J., 43

W

War dogs, 20
Westminster Pit, 56, 90, 111, 112
Westminster-Pit, The, **115**
Wild-Boar Hunt, The, 29

Willet, Jack, 73
William III, 63
William, Earl Warren, 61
Wolf, 13
Wurttemberg, Karl von, 28

X

Xerxes, 21

Y

Young Storm, 90
Young Storm and Old Storm, **93**